MARKHAM G. ROBINSON
3637 Sonoma Ave., Apt. 112
Santa Rosa, CA 95405

The Milky Way

The Harvard Books
on Astronomy

The Great Nebula near Eta Carinae. A photograph in red (H-alpha) light made with the 40-inch Boller and Chivens reflector of Siding Spring Observatory in Australia. North is toward the bottom. The estimated distance of the nebula is 2,700 parsecs. (Courtesy of Australian National University.)

The Milky Way

Bart J. Bok
and Priscilla F. Bok

Fourth Edition
Revised and Enlarged

Harvard University Press
Cambridge, Massachusetts, 1974

Contents

The Milky Way

1
Presenting the Milky Way

There is a way on high, conspicuous in the clear heavens, called the Milky Way, brilliant with its own brightness. By it the gods go to the dwelling of the great Thunderer and his royal abode. Right and left of it the halls of the illustrious gods are thronged through open doors; the humbler deities dwell further away, but here the famous and mighty inhabitants of heaven have their homes. This is the region which I might make bold to call the Palatine of the Great Sky.

Ovid, *Metamorphoses*, Book 1, lines 168–176.

In this book we invite you to join us on a brief tour along the road to the heaven of the Greeks. Modern science is providing the transportation facilities and, without its being necessary for you to leave your comfortable chair, we would like to show you the sights. Our plan is briefly as follows: We shall start off with a quiet evening at home, during which we shall get out maps and photographs of the territory that we are about to explore. We shall introduce you to some of the intrica-

cies of our celestial vehicles and then we shall get under way. First we shall pay some casual visits to the Sun's nearest neighbors, but soon we shall move on to sound the real depths of our universe. We shall visit big stars and little stars and clusters of stars within the larger Milky Way system. Between the stars we shall encounter clouds of cosmic dust and gas, some of the dust clouds so dense that they hide from our view the sights beyond. We shall, of course, linger a while on our visits to the palaces of the illustrious gods on the main road, but we shall also ask you to join us on side excursions to the places of the common people away from the well-traveled main highways.

In spite of our desire to show you all of the Milky Way system, we shall have to limit our celestial itinerary. Not infrequently along the road we shall see markers such as "Unexplored Territory," "Caution, Heavy Fog," or more encouraging signs, "Men at Work; Pass at Your Own Risk." For the Milky Way is by no means sufficiently well explored to render all of it open to celestial tourists. If you are so inclined, you may stop here and there along the road, get out your celestial Geiger counters, and do a little prospecting on your own. We hope that upon your return you will not regret having taken time for the long trip.

So, let us look at our maps and photographs and lay out the plan for the journey through the Milky Way.

The Milky Way

In most of the United States and Europe the best general view of the Milky Way can be had in the late summer on a moonless night an hour or so after sunset. The Northern Cross of the constellation Cygnus is then directly overhead, Arcturus is on its way down in the west, and in the northeast the W-shaped constellation of Cassiopeia is rising into view. If you are far from the glare of city lights and neon signs, you will have no difficulty in locating the shimmering band of the Milky Way, which can be traced through Cassiopeia and Cepheus to Cygnus and then down toward the horizon through the constellations of Aquila, Sagittarius, and Scorpius.

The Milky Way from Cassiopeia to Cygnus has the appearance of a single silvery band of varying width; but between Cygnus and Sagittarius we can distinguish two bands separated by a dark space called the Great Rift (Fig. 1).

1. The Milky Way in Cassiopeia, Cepheus, and Cygnus, a composite photograph prepared from the Palomar Sky Survey. The Andromeda Nebula, Messier 31, is shown in the lower left and the bright star Vega is near the upper right. The North America Nebula and the bright star Deneb (to the right and above the nebula) are near the center of the photograph. (From the *Atlas* prepared by Hans Vehrenberg, Treugesell Verlag, Düsseldorf.)

The western branch is quite bright in Cygnus and still discernible in Aquila, but it is lost in the wastes of Ophiuchus. The head of the Great Rift is often referred to as the Northern Coalsack. It is shown in Fig. 1 to the right of the North America Nebula. Visually, it is observed stretching to the south and west of Deneb, the brightest star in the constellation Cygnus.

There are some very conspicuous bright spots along the summer Milky Way. The star clouds of Cygnus are right overhead (Figs. 1 and 9). Though they put on a fine show, they lose out in comparison with the Cloud in Scutum—which Barnard called "the Gem of the Milky Way" (Fig. 2)—and with several bright clouds in Sagittarius. The Milky Way is still conspicuous in the constellation Cepheus, but even a cursory inspection will show that north of Cygnus it does not shine nearly so brightly as does the branch to the east of the Great Rift and south of Cygnus.

What lies beyond the horizon? Our summer night progresses. Sagittarius, Aquila, and Cygnus gradually set. As Cassiopeia rises toward the meridian other parts of the Milky Way come into view and we can follow the band through Perseus, Auriga, and Taurus;

2. The Milky Way in Aquila, Scutum, Ophiuchus, Sagittarius, and Scorpius, a composite photograph prepared from the Palomar Sky Survey. The bright star Altair is shown just below the Milky Way near the left-hand edge. The arc of Scorpius is near the right-hand edge. The Scutum Cloud is the bright star cloud to the left of the center and the Great Star Cloud of Sagittarius is to the right and slightly below the center. The connected groupings of obscuring clouds are spread in a widening band (the Great Rift) beginning as a rather narrow feature about Altair. Figure 2 is a direct continuation of Fig. 1. (From the *Atlas* prepared by Hans Vehrenberg, Treugesell Verlag, Düsseldorf.)

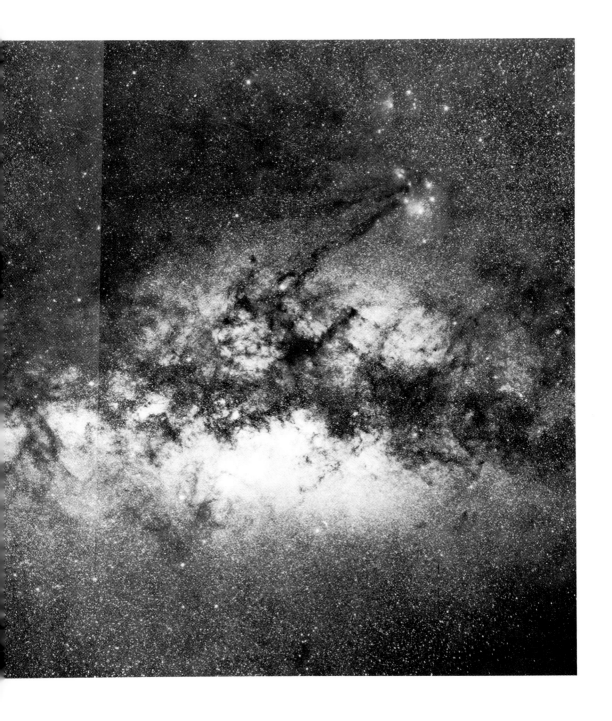

east of Taurus and Auriga it is lost in the summer dawn. But if we wait until early fall we can follow it southward through Gemini, Orion, Monoceros, and Canis Major. The Milky Way from Cygnus through Cassiopeia to Canis Major is, however, much weaker than the branches on either side of the Great Rift. In Auriga and Taurus it narrows down to a trickling little stream that is quite insignificant in comparison with the brighter sections of the summer Milky Way.

What happens to the Milky Way south of Canis Major? It is invisible from the latitudes of New York and Paris and we shall have to travel southward if we wish to see those parts. The whole Milky Way passes in review for a year-round observer in the southern tip of Florida, but for a good view we must go down to the equator or, preferably, farther south to Chile or Peru, to South Africa, or to Australia.

The section of the Milky Way from Sagittarius through Scorpius, Norma, Circinus, Centaurus, Crux (the Southern Cross), and Carina has great brilliance. In general appearance it resembles to some extent our summer Milky Way between Cygnus and Sagittarius. The star cloud in Norma is not unlike the Scutum Cloud, and the Carina Cloud appears rather similar to the Cygnus Cloud. The southern Milky Way does not show a Great Rift, such as we find from Cygnus to Sagittarius. However, it has a remarkable "dark constellation" in the connected configuration of dark nebulae stretching from the Southern Cross to Scorpius and Ophiuchus (Fig. 3). Australian aborigines referred to it as the Emu, and its ostrich-like appearance is well known to southern observers. It is best observed in the early hours of evening in July, when the Southern Cross is high in the sky and near the meridian. The Southern Coalsack represents the Emu's head, with a sharp

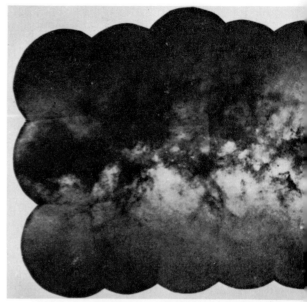

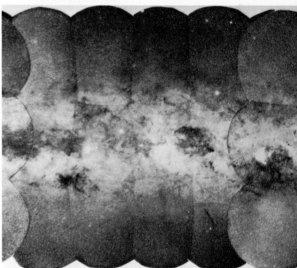

3. The southern Milky Way, a composite photograph prepared from the photographs in H-alpha (red) light reproduced in the Mount Stromlo Observatory *Atlas* by A. W. Rodgers, J. B. Whiteoak, *et al.* (Courtesy of Australian National University.)

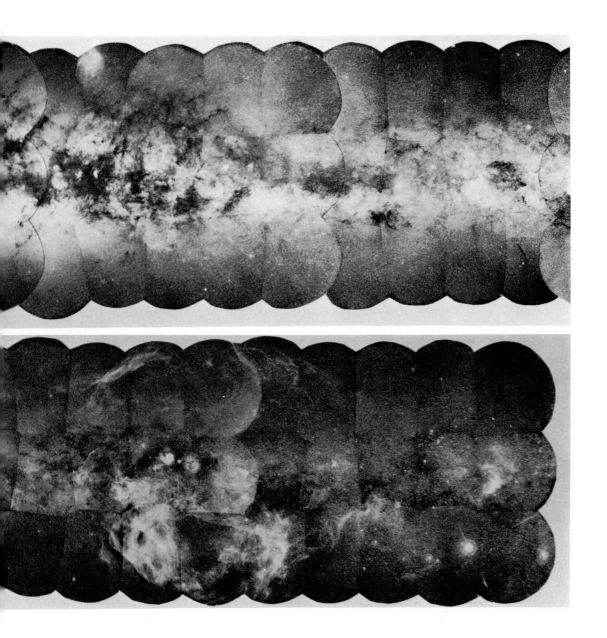

beak; the long thin neck is the narrow dark band through Centaurus to Norma, and the main body can be seen as the dark clouds in Scorpius and Sagittarius, with the thin dark legs shown by the dark lanes in Ophiuchus. The Australian aborigines can boast of having outlined the only recognized "dark constellation." It is bigger than any bright constellation, for our Emu stretches over 60° along the band of the Milky Way!

The remaining section of the southern Milky Way, which runs from Canis Major through Puppis and Vela to Carina, is in general not unlike the northern Milky Way in Cepheus and Cassiopeia. There are no marked irregularities, and a smooth band made up of thousands upon thousands of stars is clearly visible along the entire course. The whole of the band of the Milky Way forms very nearly a complete circle around the sky. We often refer to this great circle (that is, a circle that cuts the sky in half) as the *galactic circle* or *galactic equator.*

Ours are the days of large and powerful telescopes and you might well ask if there is much point to a careful naked-eye study of the appearance of the Milky Way. We earnestly believe that there is much to be learned from a survey without the use of a telescope or photographic camera. Our eyes happen to be the finest pair of wide-angle binoculars that has yet been made. A telescope is useful for the study of fine details for comparatively small sections of the sky, but no instrument is capable of revealing as well as the human eye the grand sweep of the entire Milky Way. On a good night we can directly intercompare portions of the Milky Way that are as far apart as Sagittarius and Cassiopeia. Such direct intercomparisons reveal one of the most important properties of the Milky Way, namely, that the width and the brightness of the band differ greatly from one section to another. The Milky Way attains its greatest width as well as its maximum brightness in Sagittarius. The half of the Milky Way from Cygnus through Sagittarius to Carina is generally very much brighter than the half that runs from Orion to Carina.

Telescopic Views

A good powerful pair of binoculars, or a small visual telescope, will reveal that the Milky Way is a composite effect produced by thousands upon thousands of faint stars. As we sweep across the sky with our telescope, the total number of stars in the field of view increases markedly as we approach the Milky Way. Almost two centuries ago, Sir William Herschel spent many years sweeping the sky —or, as he called it, "gauging the heavens"— with his giant reflectors. His son John later carried out the same plan for the southern sky. Their studies gave accurate data on the rates at which the numbers of stars increase toward the Milky Way. They showed that the rate of increase is very much larger for the fainter than for the brighter stars. If we use a 3-inch telescope and compare two fields, one in the Milky Way and the other in a direction at right angles to it, near one of the so-called galactic poles, then we count three to four times as many stars in the Milky Way field as near the pole. If we repeat the experiment with a 12-inch telescope the ratio is nearly ten to one.

Among the celestial objects that delight the amateur with a modest telescope of his own are the clusters of stars, the open clusters as well as the globular clusters (Figs. 11 and 12), and the beautiful nebulae. How are these prize celestial objects distributed relative to the Milky Way? Star clusters generally show a preference for the Milky Way. The open

4. The Milky Way in Sagittarius. (From a photograph taken by Ross with a 5-inch camera at the
Lowell Observatory.)

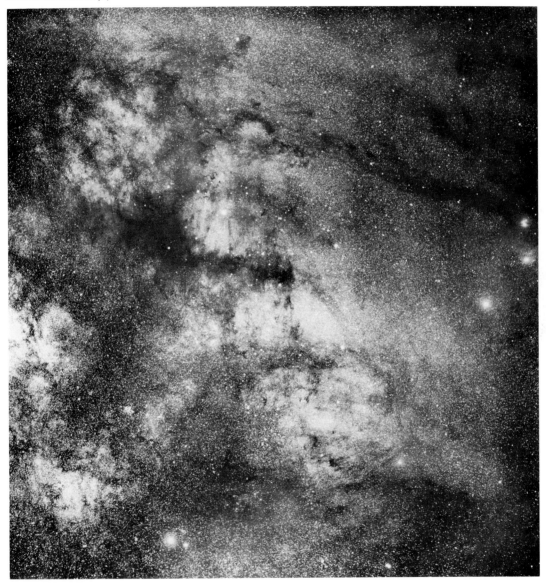

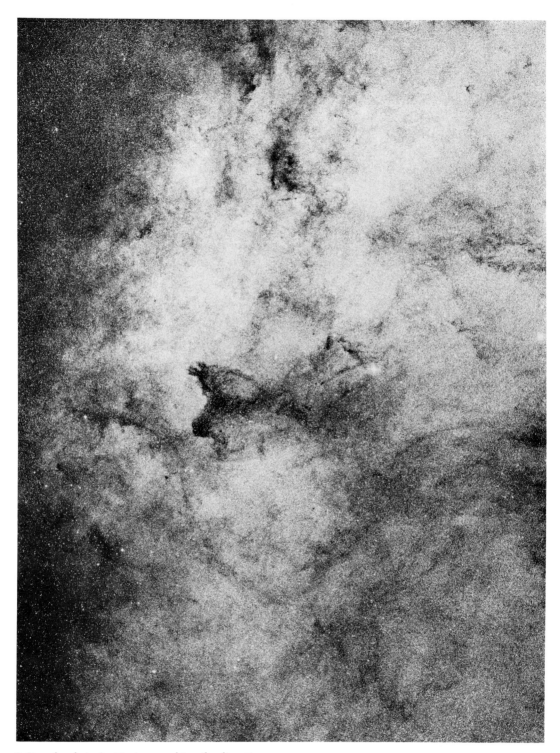

5. Star clouds in Sagittarius, marking the direction
toward one edge of the nucleus of our Milky Way
system. North is at the top; west is to the left.
(48-inch Palomar Schmidt photograph.)

6. The Great Rift near Altair. The photograph by
Ross shows the two star-rich branches of the
Milky Way near the bright star Altair (lower left-
hand corner) bordering the dark nebulae that
mark a section of the Great Rift in the Milky Way.

7. The Great Star Cloud in Scutum, Barnard's
"Gem of the Milky Way." (From a photograph by
Barnard made at Mount Wilson Observatory.)

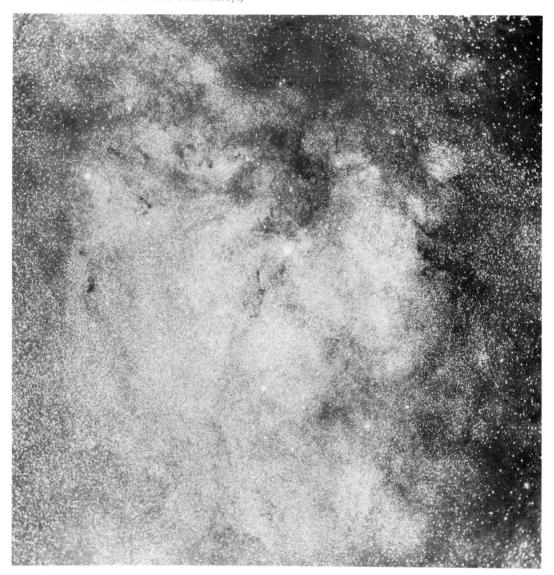

8. A region north of Theta Ophiuchi. A network of dark nebulosity overlies this rich star field. Note the remarkable snakelike formation in the lower part of the photograph and the small roundish dark nebulae near it. (From a photograph by Barnard.)

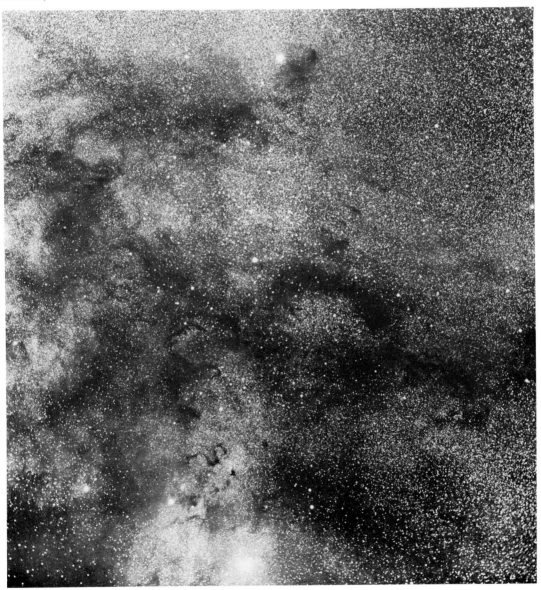

9. Bright and dark nebulosities near Gamma
Cygni. (48-inch Palomar Schmidt photograph.)

10. Concentration of stars toward the Milky Way. The Milky Way in Cygnus, very rich in stars, and a small section of the Great Rift are shown in the lower left-hand corner. The photograph shows the manner in which the faint stars gradually thin out as we pass in the sky farther from the central band of the Milky Way. (From a photograph by Ross with a 5-inch camera.)

11. The galactic cluster Kappa Crucis, the "Jewel Box" of the southern Milky Way. (60-inch Rockefeller reflector, Boyden Observatory.)

clusters, of which the Pleiades, the Hyades, Praesepe, and the double cluster in Perseus are the prototypes, are mostly found close to the band of the Milky Way; faint open clusters lie almost without exception within a few degrees of the Milky Way circle. The globular clusters, such as the well-known one in Hercules, appear to prefer the regions between 5° and 20° from the Milky Way circle.

Later in this book we shall see that there are striking differences in the numbers of stars in open and in globular clusters. For the present, we shall say that the open clusters are those with somewhere between 20 and 2,000 members, whereas globular clusters may have anywhere from 10,000 to possibly 1,000,000 members. Most of the relatively nearby open clusters are readily resolved in a fairly small telescope, but it takes a large instrument to see individual stars in globular

clusters. The name "open cluster" comes directly from the appearance of the cluster in the telescope.

The behavior of the nebulae may at first seem puzzling. Some of the finest nebulae, the Carina Nebula (Fig. 13) and the Orion Nebula (Fig. 70), for instance, are either in the band of the Milky Way or close to it; but there are others—especially those with spiral structure—that seem to avoid the Milky Way and occur mostly at some distance from the central band. The puzzle is easily explained. There are really two completely different varieties of objects called "nebulae"—the diffuse or gaseous nebulae, which are part and parcel of our Milky Way, and the spiral nebulae. The latter are not nebulae at all but are systems of stars in their own right, called *galaxies*, all of them far beyond the distances with which we shall concern ourselves in the pres-

12. The globular cluster Omega Centauri. (An enlargement from a photograph in red light made with the Armagh-Dunsink-Harvard telescope of the Boyden Observatory.)

ent volume. Here we limit ourselves to the study of the Milky Way system, our Home Galaxy, so to speak.

There is one further property of the globular clusters that will certainly be noted by a thorough observer with a visual telescope. He finds many globular clusters when our northern "summer" Milky Way is around, but he cannot observe many during the winter when Capella is high in the sky. This observer soon comes to the conclusion that the globular clusters in their own peculiar way favor one half of the Milky Way. They appear to be particularly fond of the region around Sagittarius, where one third of all known globular clusters are found in an area covering scarcely 2 percent of the entire sky. Open clusters and diffuse nebulae are spread more evenly along the band of the Milky Way.

Photographic Appearance

Purely visual inspection and measurement play at present a very minor role in Milky Way research. Photographic techniques came into general use about the turn of the century, and to these we have added the more recent techniques of photoelectric and radio research. When it comes to introducing beginners to the intricacies and beauties of the Milky Way, there is no substitute for photographs made with modern photographic telescopes.

Let us start with the map of the southern Milky Way that is shown in Fig. 3. It has been made by matching, cutting, and pasting together a series of black-and-white prints from photographic negatives made with an 8-inch, $f/1$, Schmidt camera at Mount Stromlo

13. The Carina Nebula. The emission nebulosity
in the region of the Carina Nebula covers a large
area of the sky, about 25 times the area subtended
by the full Moon. The photograph was made
with the 24/36-inch Curtis-Schmidt telescope of
Cerro Tololo Inter-American Observatory. The
frontispiece shows the section with densest
nebulosity.

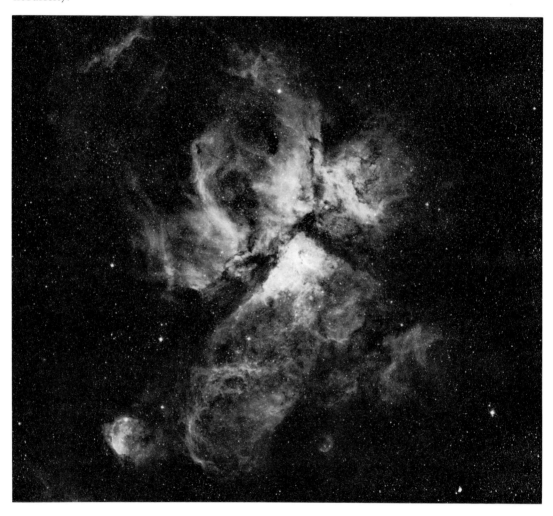

Observatory, Australia. The upper photograph covers the half circle from the star clouds in Scutum and Sagittarius, on the left, to the Coalsack and the Southern Cross. The lower photograph carries us past the Coalsack and the Carina Nebula to Canis Major and Monoceros on the right; the bright star in the lower right-hand corner is Sirius.

In spite of its small scale, the composite map shows clearly some of the important features to which we have already referred. In some places the resolution is not enough to show the individual stars, but in general we find here direct confirmation of the stellar nature of the Milky Way. The changing appearance that is noted as we proceed from one photograph to the next is not caused by atmospheric difficulties or differences in exposure time, but is mostly the result of true variations in brightness along the Milky Way. The Milky Way is very much more spectacular in Sagittarius than near Sirius.

The small-scale photographs used in the preparation of the composite map were made through a special red filter and on a red-sensitive photographic emulsion. This was done to show not only the stars of the southern Milky Way but also the gaseous nebulae that are present. These nebulae radiate profusely in the red light of the hydrogen line, H-alpha, and, by the use of the proper combination of color filter and photographic emulsion, they become very apparent. One of the finest large nebular shells is shown in Fig. 3 in the lower panel, to the left of the image of Sirius. It is named the Gum Nebula after the late Colin S. Gum, who, as a young astronomer, pioneered in filter photography of the southern Milky Way with wide-angle cameras. He was the first astronomer to draw attention to this magnificent nebula.

When we turn to detailed studies of the Milky Way, we need both a more open scale and greater penetrating power than that given by our small-aperture cameras of short focal ratio. Here is where the Schmidt-type telescopes with apertures of 24 inches or more and the large reflectors enter the picture. In the frontispiece and in Fig. 13 we show two aspects of the famous nebula near Eta Carinae, truly the most beautiful diffuse nebula of the Milky Way; because of its far southern position it is not visible from northern latitudes. Figure 13 shows the field of the nebula and its surroundings as photographed with the Curtis-Schmidt Telescope of the University of Michigan at the Cerro Tololo Inter-American Observatory in Chile. The frontispiece is an enlargement of a small portion of the nebula; here we are impressed with the grand sweep of the swirling gases and the intricate dark patterns overlying them.

The photographs in our book show principally the spectacular star-rich sections directly along the band of the Milky Way. The views are less striking when we consider photographs of fields at some distance from the galactic circle. By comparing photographs of different star fields made with the same telescope and identical exposure times, we find that the average numbers of stars per unit area of the sky are greater for fields close to or in the band of the Milky Way than for fields at some distance from the galactic circle. The larger the telescope and the longer the exposure times, the more striking becomes this contrast between counted numbers of stars for fields close to and far from the galactic circle. This observation alone suggests that the visible phenomenon of the Milky Way has great depth and that our Sun is located near the central plane of a vast star system that is highly flattened.

Our Milky Way System: A Model

If this were to be a detective story we might wish to present first all available evidence and then hide the solution in some uncut pages toward the end of the book. Our story is not so simple. The evidence is so incomplete in spots that we are nowhere near the final solution of the Milky Way mystery. Under these circumstances we might as well give away our "secrets" at the start. We shall make reading easier by providing a model of the Milky Way system. The descriptive material of the preceding pages provides a basis for such a model.

Visual as well as photographic counts show that the faintest stars are relatively most concentrated toward the band of the Milky Way. Since, on the average, the fainter stars are the more distant ones, we have thus direct proof that our Milky Way outlines a flattened system of stars. The Milky Way has great depth. Some of the stars that contribute to the Milky Way phenomenon may be only a few hundred parsecs away, but others are at distances of several thousands of parsecs from our Sun (1 parsec is equivalent to 3.26 light-years, about 20 trillion miles or 30 trillion kilometers). Since the Milky Way appears as a great band encircling the sky, cutting it into nearly equal parts, the Sun must be located close to the central plane of the system. Is it located anywhere near the center of the system? For many centuries astronomers believed this to be so, but if you have read the preceding pages carefully you will have found some indications that the Sun is located far from the center of our Galaxy. One of the most striking visual features of the Milky Way is that the half centered upon the stars of Sagittarius is wider and more brilliant than the part in Orion, Taurus, and Auriga. This means that the center of the Galaxy probably lies in the direction of Sagittarius.

There is much evidence to show that the galactic center lies in the direction of Sagittarius. The globular clusters exhibit a very pronounced concentration toward the Sagittarius region. Harlow Shapley showed that the globular clusters are among the most distant observable galactic objects; their irregular distribution is strong evidence of the existence of a distant center in Sagittarius. The concentration toward the Sagittarius region is shared by many types of objects that can be observed at great distances from the Sun, such as new stars, or novae, distant variable stars, and planetary nebulae. There is the evidence from galactic rotation, which postulates the existence of a distant center in the general direction of the Sagittarius clouds, and, finally, the indication that the strongest radio radiations come from precisely the same part of the sky.

So far we have given no indication with regard to the approximate dimensions of our Milky Way system. The evidence on this point will be presented in due course in later chapters, but we might as well complete the description of our model now. There is fairly general agreement among astronomers that the center of our Milky Way system lies somewhere between 8,000 and 11,000 parsecs from the Sun, with 10,000 parsecs representing the best compromise value.

The model is shown diagramatically in Fig. 14. There is no such thing as a sharply defined outer boundary to the Milky Way system. The solid line in our diagram is drawn through points where the space density of the stars will hardly exceed a few percent of the value near the Sun. In later chapters we shall read occasionally about varieties of stars beyond these boundaries, stars that still belong quite definitely to our Milky Way system.

One of the main problems of Milky Way astronomy is to obtain information regarding

14. A schematic model of our Galaxy. The boundaries shown outline the parts of our Milky Way system inside which the majority of the stars are located. The over-all diameter of the Galaxy in the central plane is of the order of 30,000 parsecs. Our Sun occupies a position close to the central plane at a distance of 10,000 parsecs from the center.

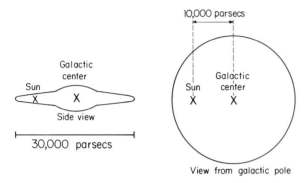

the arrangement of stars, gas, and dust near the central plane. Astronomers 50 years ago guessed that our system is probably a spiral system, and this is now well established. We shall find that, superposed on a fairly smooth conglomerate distribution of rather unspectacular stars, there exists a system of spiral arms. The spiral features can be traced rather well through the plotting of the positions of the blue-white supergiant stars and of the regions of greatest density of the interstellar gas and dust.

What about the motions in the Milky Way system? The system is apparently revolving at a terrific pace around the distant center. This is hardly surprising, for the system could not possibly stay as flat as it appears to be without a rapid rotation in the general plane of symmetry. The rate of rotation is fast; our Sun is whirled around at a speed of approximately

250 kilometers per second. That ought to be just about fast enough to suit our readers and should provide enough momentum to propel them through the basic fact-finding chapters that follow, onward into the realms of fancy.

Terms and Concepts

Each science develops its own special language, which is useful and necessary for communication but which often seems baffling to the uninitiated. To clear the atmosphere, we shall first write briefly in this section about some of the terms and concepts that all prospective students of the Milky Way should know before they delve into the field.

Milky Way and Galaxy. The physical phenomenon that we see in the sky is generally referred to either as the *Band of the Milky Way* or simply as the *Milky Way*. It stretches around the sky as very nearly a great circle

15. The great spiral in Andromeda. The spiral galaxy in Andromeda bears in all likelihood considerable resemblance to our own Milky Way system. Two companion ellipsoidal galaxies are also shown in the photograph. (48-inch Palomar Schmidt photograph.)

and it marks the central plane of our Milky Way system, shown diagrammatically in Fig. 14. Our Milky Way system, with its some 100 billion stars and with its content of interstellar gas and dust, is the stellar system to which our Sun—just another star—belongs.

There are millions of other systems not unlike ours in the observable universe, and we call these *galaxies*. In other words, our own Milky Way is just one of many galaxies and we can think of it as our Home Galaxy. Many galaxies, including our own, show spiral structure and these are spoken of as *spiral galaxies*, in contrast to the more amorphous galaxies such as the *ellipsoidal* and the *irregular galaxies*. To all but the few visual observers with very large telescopes, the spiral galaxies look quite nebulous, and visually their spiral patterns hardly reveal themselves. For that reason spiral galaxies are often called *spiral nebulae* (a name that is really out of date and wrong), and the other varieties are similarly called ellipsoidal and irregular nebulae; owners of small telescopes like to show their friends the Andromeda Nebula, which is really a spiral galaxy (Fig. 15). The only true nebulae are the gaseous nebulae, like those in Orion and Carina, or the dust nebulae like the one associated with the Pleiades cluster, or the Southern Coalsack.

The first list of clusters, nebulae, and galaxies was prepared a century and a half ago by a French astronomer, Charles Messier, and most of the brighter objects are still usually known by their *Messier numbers*. For instance, the Andromeda Nebula, a spiral galaxy, is Messier 31 and the Orion Nebula is Messier 42. Astronomers often have occasion to refer to extensive lists of clusters, nebulae, and galaxies prepared by Dreyer, the *New General Catalogue* (NGC) and its successor, the *Index Catalogue* (IC); Messier 31 then

becomes NGC 221. The Messier numbers suffice as a rule for our purposes.

Magellanic Clouds. Our Milky Way system is accompanied in space by two smaller systems, the *Large* and the *Small Magellanic Clouds* (Fig. 17). Another of the Harvard Books on Astronomy, Shapley's *Galaxies*, deals with these neighboring galaxies, which tag along with our Galaxy very much in the manner in which a pair of moons accompanies a planet.

Galactic Latitude and Longitude. In order to fix the positions of stars and other celestial objects in the sky, we imagine a celestial sphere of very great radius and concentric with the Earth (Fig. 18). By extending the Earth's axis of rotation skyward, we pierce the imaginary sphere at two points called the *north* and *south celestial poles*. The north celestial pole will be directly overhead for an observer at the North Pole on the Earth. Next, we define the great circle halfway between the two poles and call it the *celestial equator*. The *vernal equinox* is then defined as the point on the celestial equator where the Sun crosses it at the beginning of our northern spring. The position of any star or other celestial object is then given by its *right ascension* and *declination*. The declination measures how far in degrees on the celestial sphere the object is north or south of the celestial equator; the right ascension measures (along the celestial equator, in hours, minutes, and seconds—$1^h = 15°$) how far it is east of the vernal equinox. Since the positions of the celestial poles, the equator, and the vernal equinox are subject to slow progressive changes, the right ascension and declination of a given object will not remain precisely constant with time and we refer all right ascensions and declinations to a given *epoch*, say 1900, 1950, or 2000.

16. The spiral galaxy Messier 101 in Ursa Major. A photograph made with the 200-inch Hale reflector. (Hale Observatories photograph.)

17. The Large Magellanic Cloud. A companion galaxy to our own Milky Way system. (ADH photograph, Boyden Observatory.)

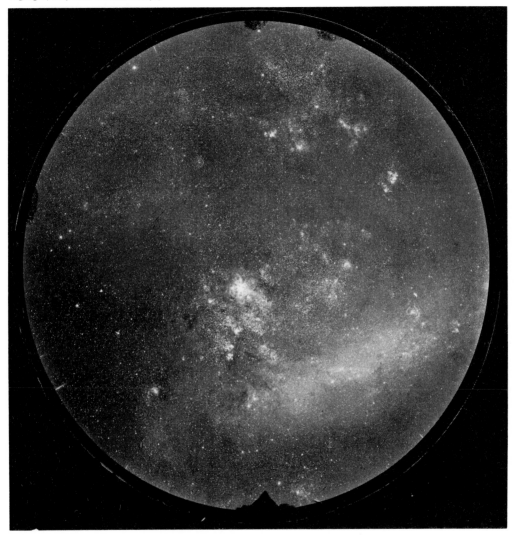

18. The celestial sphere. The diagram helps to visualize right ascension and declination. The approximate right ascension of the star shown is 75° (5 hours) and the declination is 40°N.

19. Galactic latitude and longitude. The star shown has approximately a galactic longitude $l = 100°$ and a galactic latitude $b = +20°$; its right ascension is about 20^h and its declination is approximately 70°N.

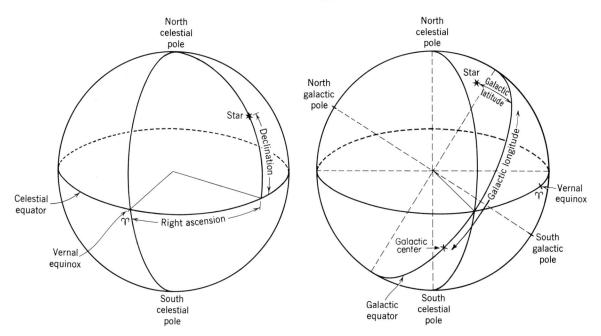

The equatorial system of defining positions in the sky is the basic one, but for studies of the Milky Way we often prefer the use of a special system, better adapted to our work. We noted that the band of the Milky Way follows roughly a great circle in the sky. For purposes of research we draw a great circle through the band of the Milky Way as it appears to us and this circle we call the *galactic equator* (Fig. 19). The points 90° away from the galactic equator on either side of it are called the *north* and *south galactic poles*. The position of any object in the sky can then be described by its *galactic latitude*, measured in degrees north or south from the

galactic equator, and its *galactic longitude*, measured eastward along the galactic equator. The starting point, or zero, of galactic longitude has been established by international convention as the point on the galactic equator that marks the direction toward the galactic center as seen from the Sun. The galactic equator is inclined roughly 62° with respect to the celestial equator. The position of the galactic center (1950 epoch) is $17^h42^m37^s$ (right ascension), $-28°57'$ (declination). The north galactic pole is located (1950 epoch) at $12^h49^m02^s$ (right ascension), $+27°23'$ (declination).

20. The Steward Observatory station on Kitt Peak in Arizona. The dome of the 90-inch reflector is shown on the right, that of the 36-inch reflector on the left. The first optical pulsar was discovered with specialized electronic equipment attached to the 36-inch reflector. (University of Arizona photograph.)

Telescopes and Auxiliary Equipment

Telescopes of many varieties are the basic tools of the astronomer. For Milky Way research we need the best and most powerful telescopes available. It may be helpful to remind the reader that there is a first broad distinction between *refractors* (telescopes with lenses) and *reflectors* (telescopes with paraboloidal mirrors). For Milky Way photography, multiple-lens refractors have proved useful, especially those that permit photography with excellent definition of a large area of the sky on one single photograph; these photographic refractors are often called cameras, even though their aperture (the diameter of the primary lens) may be 15 to 20 inches. The largest telescopes in use now are the reflectors, in which the light is gathered by a carefully ground and polished paraboloidal mirror covered with a thin aluminum coating. These instruments are our most powerful light collectors, but they suffer from having relatively small areas of perfect focus.

The development of the *Schmidt-type telescope* (Fig. 24) proved a great boon to Milky

21. The Steward Observatory 90-inch reflector on
Kitt Peak in Arizona. (University of Arizona photograph.)

22. The domes of the 150-inch reflector and of the 24/36-inch Curtis-Schmidt telescope of the University of Michigan at Cerro Tololo Inter-American Observatory in Chile. (Photograph by the authors.)

23. The 158-inch Mayall reflector of Kitt Peak National Observatory in Arizona. (KPNO photograph.)

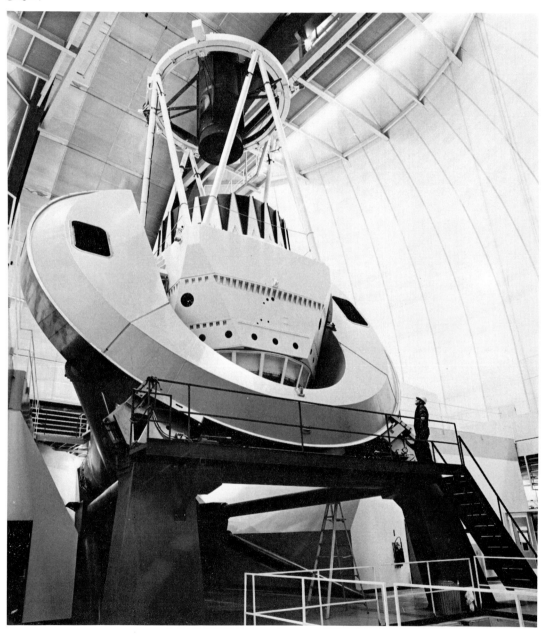

24. The Armagh-Dunsink-Harvard (ADH) tele-
scope at the Boyden Observatory, which is of
Baker-Schmidt design. (Photograph by *The Friend*,
Bloemfontein, South Africa.)

Way research. This telescope consists of a primary spherical mirror, with a correcting lens
placed near the center of curvature of the
mirror. This combination has excellent focus
over a large area of the sky, together with the
great light-gathering power of the large primary mirror. Modifications, such as the
Baker-Schmidt, are extensions of the basic
principle enunciated in 1929 by Schmidt of
Hamburg. The most famous of all is the 48-
inch Palomar Schmidt (Fig. 25), which has
been used for the photography in blue and in
red light of the entire sky within reach from
Palomar Mountain in southern California.
The *National Geographic–Palomar Sky Survey* contains 1,000 prints of the blue-sensitive
photographs and 1,000 prints of the photographs in red light. Each print covers an area
of the sky equal to 6° × 6°. No observatory
library is considered complete without a set
of the *Sky Survey* prints or, preferably, a set
of prints on glass.

25. The 48-inch Schmidt telescope at Mount Palomar, which has completed the National Geographic–Palomar Survey.

A variety of telescope that is very important to Milky Way research is the *radio telescope*. In its simplest form it is the paraboloidal reflector of optical astronomy scaled up to the dimension needed for radio studies of the Milky Way. The 100-meter steerable paraboloid radio telescope near Bonn in Germany and the precision 140-foot and the 300-foot radio telescopes at the National Radio Astronomy Observatory in Green Bank, West Virginia (Figs. 26 and 27) are among the largest Northern Hemisphere instruments used for galactic research. The 210-foot precision radio telescope at Parkes in New South Wales, Australia, has long dominated research from southern latitudes.

We shall describe in the chapters that follow several of the more useful auxiliary tools of the Milky Way astronomer, but at this early stage a few deserve to be mentioned briefly. Success in direct Milky Way photography depends wholly on the quality (speed, graininess, color sensitivity) of the *photographic emulsions* provided by the manufacturer and on the characteristics of the special *color filters* employed for each particular project. For studies involving spectra of single stars, or of star fields, we require auxiliary *spectrographic equipment.* The basic instrument is either a relatively small spectrograph used at the Newtonian or Cassegrainian focus of a large reflector, or—for highest attainable spectral dispersion—a coudé spectrograph. Photographic recording of spectra is being supplemented by automatic photoelectric recording. Photography is being assisted by image-strengthening techniques, in which use is made of the image-conversion approach. For precision work on the measurement of brightnesses and colors of stars, the astronomer depends now mostly on the *photoelectric photometer.* Basic standards are es-

tablished by photoelectric techniques and the photographic plate serves as a useful tool for extending the work to many stars, using the photoelectric standards as a basis. Photographic and especially photoelectric and spectrographic work is being extended beyond the traditional violet-blue-green-yellow-orange-red parts of the spectrum. Space research probes the Milky Way beyond the near ultraviolet into the regions of shorter wavelength, all the way to the x-ray region. In recent years great advances have been made in the infrared parts of the spectrum, so much so that the earlier gap between the near infrared and the radio region has practically been filled.

Instrumental advances on all fronts are assisting further developments in galactic research. Our modern large reflectors are now often fitted with primary mirrors that have a reflecting surface slightly different from the traditional paraboloidal shape. When used in conjunction with specially designed secondary mirrors, these Cassegrain combinations provide fields for photography with wide-open scales on the photographs. One photograph can cover with near-perfect definition areas of the sky with diameters of the order of 1° or more.

Milky Way research, like all astronomy— and for that matter all modern science—is profiting tremendously from the worldwide developments of automatization and data processing. Large-scale undertakings for the measurement of hundreds of thousands of stellar brightnesses, colors, or proper motions are no longer considered beyond the realm of possibility.

The Milky Way astronomer must pay careful attention to his basic tools and to the methods available for their proper and effective use. The quality of our research depends on the accuracy and proper use of our tools.

26. The radio telescopes of the National Radio Astronomy Observatory at Greenbank, West Virginia. The three telescopes on the right represent the interferometer array used for the detailed mapping of radio sources. On the left in the foreground is the 300-foot altazimuth radio telescope, steerable only in altitude, which has been used very effectively for 21-centimeter research (see Chapters 8 and 10); it is now housed inside a dome transparent to radio waves. In the back is shown the 140-foot radio telescope (see Fig. 27). (Courtesy of the National Radio Astronomy Observatory.)

27. The 140-foot steerable radio telescope of the National Radio Astronomy in Greenbank, West Virginia. (Courtesy of the National Radio Astronomy Observatory.)

2
The Data
of
Observation

Our eyes are our first tools. The sky lies open before us, for astronomy is an observational rather than an experimental science. But our eyes are weak and forgetful and therefore we build telescopes that can gather more light, cameras that can store lasting impressions, prisms and spectrographs to disperse the light into spectra, and photometers that are more sensitive than our eyes and can measure stellar brightnesses and colors with precision.

Our eyes are sensitive only to the part of the spectrum from the blue-violet to the red. Figure 28 shows the total range of wavelengths that can be detected and studied with modern techniques for gathering and recording light. Visible light is one form of electromagnetic radiation that is observed. Each electromagnetic wave is characterized by its wavelength or by its frequency of vibration. The wavelengths of light in the visual, ultraviolet, and infrared rays are measured in *angstrom units*, 1 angstrom unit being equal to 1 one-hundred-millionth of a centimeter (10^{-8} cm). But for radio radiations it may be more convenient to measure wavelengths instead in millimeters, in centimeters, or even in kilometers! Astronomers working in the radio or infrared parts of the spectrum often prefer to use the frequency of radiation instead of the wavelength. Since the wavelength λ (lambda) times the frequency ν (nu) equals the velocity of light c, we have:

$$\nu = \frac{c}{\lambda}.$$

A frequency of 1 cycle per second is called 1 *hertz/Hz*, so named in honor of the great German physicist Heinrich Hertz. Radio astronomers measure their frequencies mostly in *megahertz*; 1 megahertz (MHz) = 1 million Hz.

Figure 28 shows the ranges in wavelength and in frequency for the varieties of radiation that can now be observed. Beyond the violet or short-wave length end of the blue rays are

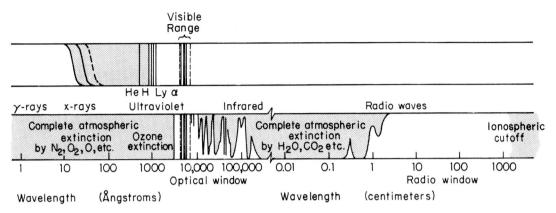

28. The spectrum of electromagnetic waves from gamma rays to long radio waves. In the lower strip, the optical and radio "windows" are indicated by white areas and the regions of atmospheric extinction by shaded areas. The upper strip shows the range of radiations accessible to detectors flown in a rocket or satellite above the Earth's atmosphere. Absorption by hydrogen, and to some extent by helium, cuts off the light of distant stars in the ultraviolet far beyond the cutoff of the Earth's atmosphere. Note, however, that gamma rays, most x-rays, and other regions of the spectrum can be observed without much interference. The narrow range of wavelengths to which the eye is sensitive is indicated. (From Lawrence H. Aller, *Atoms, Stars, and Nebulae*, revised edition, 1971.)

the invisible ultraviolet rays, the x-rays, and the gamma rays; beyond the red are the near-infrared heat rays and beyond these the far infrared and the radio microwaves. The radiation outside the visual range can be studied with the aid of the photographic plate and with special receiving devices, such as the photoelectric cell for the blue-violet. We use special infrared-sensitive cells, bolometers and thermocouples, for the infrared, and radio telescopes for the radio range. Earth satellites are helpful in carrying our instruments beyond the Earth's atmosphere. They have especially enlarged our knowledge of the far-ultraviolet radiation and of x-rays reaching us from the stars and nebulae of our Galaxy.

Stellar Brightnesses

One of the first questions that we shall want to answer is, How many stars are there?

But, even if we could answer this simple question, we would immediately want to know how many there are in each successive class of brightness. At once we have met with one of the most difficult problems of the modern astronomer: How are we to measure accurately the amount of light that reaches us from individual stars?

Since the human eye was the first instrument used for observing the stars, we inquire first into the simple rules that govern visual estimates of brightness. Our appreciation of the difference in brightness of two lights is always a relative rather than an absolute matter. Two faint stars may seem to differ appreciably in brightness, but two bright stars that differ by the same absolute amount would appear almost identical, since in the latter case the difference would be a negligible fraction of the total stimulus that we receive. In other words, our eyes estimate ratios of brightness rather than absolute differences.

Hipparchus was the first to classify the naked-eye stars according to brightness. Later it was agreed that his six "magnitude" classes should be so taken that a standard first-magnitude star (the average of the 20 brightest stars in the sky) gives us 100 times as much light as a star of the sixth magnitude, which is about the limit of vision for the human eye. Since we are interested in *ratios* of brightness, we define a difference of 1 magnitude as corresponding to a ratio equal to the fifth root of 100, which is 2.512. A difference of 2 magnitudes corresponds to $2.512^2 = 6.31$, 3 magnitudes to $2.512^3 = 15.85$, 4 magnitudes to $2.512^4 = 39.82$, and 5 magnitudes to $2.512^5 = 100$.

We need perhaps to stop for a moment and consider how powerful a geometric factor of this kind can be. As viewed from the Earth, a sixth-magnitude star emits 1/100 the light of a first-magnitude star; an eleventh-magnitude star emits only 1/10,000 the light of a first-magnitude star. With the 200-inch telescope, stars of the twenty-third magnitude are recorded and measured. Such a star has $1/2.512^{22}$ or 1/630,000,000 of the light of a first-magnitude star. Please note that the faintest stars have the greatest numerical magnitudes! We start counting with magnitude zero, stars like Vega or Capella; a star like Sirius, which is brighter than Vega, is then given a negative magnitude, in this case -1.5.

Photoelectric Photometry

The instruments used for measuring with precision the brightnesses and colors of stars are called *photometers*. Among the photometers designed to surpass the eye in sensitivity and accuracy, the photoelectric photometer reigns supreme. Its accuracy is unsurpassed and it is an instrument that is relatively simple to use. The light from the star strikes a sensitive photoelectric tube, or photocell, and releases electrons, which constitute a very small current. This electric current is directly proportional to the intensity of the light falling on the photocell. The electrons released by the impinging starlight are accelerated by an electric field in the photocell and strike a second sensitive surface, where each original electron may release two or more additional electrons. This process is called *photomultiplication* and the electronic tubes that incorporate this feature are called *photomultiplier tubes;* the photo tube in standard use is referred to as the IP 21 photomultiplier cell. In most of the modern photomultiplier tubes used for astronomical work there are as many as nine successive stages of photomultiplication built into the tube so that amplication by factors of the order of 1,000,000 is achieved. The current produced by the photo tube is passed through an amplifier, and in the older instruments it is then recorded by the deflection of a pen moving across a roll of paper. Figure 29 shows schematically the type of record produced by a photoelectric photometer. In modern equipment the chart recording is dispensed with and the amplified photocurrents are numerically recorded in a form suitable for subsequent computer reduction.

Thermally excited currents are present in all photomultiplier tubes, even though no light is falling on the photosensitive surface. As a measure of the photocurrent produced by the light of a star we take the difference between the reading on our record when the star's light is striking the photosensitive surface and the reading when no light falls on the cell. Figure 29 shows how measurements

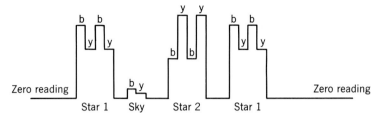

Zero reading   Zero reading

Star 1   Sky   Star 2   Star 1

29. A schematic photoelectric tracing. The photo-electric photometer has usually one or more filters. In this schematic drawing the star is first observed in blue light, then in yellow light, and the observations are then repeated. After the first star is observed, a reading is taken on the sky where no star is visible, also in both colors. Then follow observations of star 2 and to close the sequence a repeat set on star 1. From the curves one would obtain the ratio of brightness of the two stars (equal to differences of magnitude) in blue and yellow light. The first star is brighter in blue light than in yellow light; the second is brighter in yellow light.

of the "zero" or "dark" current are made at regular intervals throughout the period of observation. The scale of brightness ratios or magnitudes is provided directly by the recording, and standard stars are observed regularly for the dual purpose of relating magnitudes to an agreed-upon zero point and of eliminating effects of changing atmospheric extinction.

Also included in the observation (Fig. 29) are so-called "sky readings." In spite of its apparent blackness, the sky background contributes a measurable brightness to the recorded deflections. Most of the radiation from the sky background is excluded from the measurements by observing the star through a small diaphragm at the focal plane of the telescope, but a little bit of sky always succeeds in peering through the diaphragm along with the star. To eliminate this sky effect, we take two readings in succession, the first of the star with the background radiation included, the second of a neighboring spot in the sky without the star. The difference between the two deflections measures the contribution from the star alone. Along the band of the Milky Way special problems arise when we attempt to measure faint stars, since it is not easy to obtain nearby sky readings free from intruding stars.

One of the advantages of the photoelectric photometer is the fact that it measures the amount of light received from the star in a direct or linear fashion: twice as much energy received by the instrument gives twice the deflection on the record. Photoelectric accuracy can be very high; it is possible to be sure of the brightness of any star to within two or three thousandths of a magnitude. Modern photoelectric methods were originally developed by Stebbins and Whitford at the University of Wisconsin and at the Mount Wilson Observatory. Photoelectric photometers are now standard equipment at practically all observatories, north and south.

Baum, at the Mount Wilson and Palomar Observatories was the first astronomer to succeed in measuring photoelectrically, with the aid of the 200-inch Hale telescope, the brightness of a twenty-third-magnitude star. To do so required special preparations and the use of techniques of photoelectric measurement that differ somewhat from those we have described, for the star in question is far too faint to be seen by the observer at the telescope and hence special ways of setting the telescope must be employed. First, a photograph of the region is taken and the difference of position between the faint star and a nearby bright star is measured. The photometer is mounted on a base that is attached to the telescope. Accurate linear scales make it possible to shift the photometer precisely and by known amounts sidewise or up or down. First the brighter of the two stars is centered on the diaphragm and then the photometer is shifted by exactly the amounts indicated by the photographic measurements. The faint star should now be centered precisely on the diaphragm and its magnitude can be determined by the customary techniques. For such very faint stars, we do not measure the instantaneous electric current produced by the very weak light of the star; instead we count the total number of light quanta, or photons, captured in the course of a few minutes to several hours and compare this number with the number of photons counted in the same interval for a somewhat brighter star of known magnitude.

Stellar Colors

The color characteristics of a system of magnitudes is determined by the range of wavelengths recorded by the receiver, be it a photocell or a photographic plate, and by the color transmissions of the filters that are used. One of the systems of magnitudes now much in use is the $UBVRI$ standard system, first recommended and developed by H. L. Johnson. In this system, ultraviolet (U) magnitudes are obtained when a special ultraviolet filter is placed in the light path from the star to the photomultiplier cell. This filter has maximum transmission at a wavelength of 0.35 micron (1 micron = 0.001 millimeter), and it has a passband—that is, a range of

wavelengths transmitted—with a width of 0.07 micron. To obtain blue (B) magnitudes, we use a standard filter with maximum transmission at 0.43 micron and with a passband width close to 0.10 micron; for visual (V) magnitudes the center of transmission by the filter is at 0.55 micron and the width of the passband is 0.08 micron. The two standard colors for the infrared have been named R (centered at 0.7 micron) and I (centered at 0.9 micron). The $UBVRI$ system has been accepted by the International Astronomical Union as *the* standard system of magnitudes. Certain stars, specifically named, define the zero point from which we count magnitudes in each selected color and the measured magnitudes of all other stars are referred to these agreed-upon standards.

The color of a particular star is given by its *color index*. For example, the blue-visual color index of a star is defined as the difference between the star's blue (B) and visual (V) magnitudes; it is given as $B - V$. Similarly, one may define for each star the color indices $U - B$, $V - R$, and $R - I$.

Starlight is cut-off by the earth's atmosphere at 0.3 micron, and the cutoff is so sharp and strong that there is no chance of extending the standard system to wavelengths shorter than 0.3 micron, at least for studies from Earth-based observatories. The situation is different for observations from satellites. Extensive research is now in progress based on satellite observation. As of now, it seems that even from space observatories we may have relatively little opportunity for research at wavelengths shorter than 0.09 micron, principally because of the galactic fog that is produced by the neutral hydrogen atoms of the interstellar gas; these atoms absorb most of the far-ultraviolet radiation beyond the Lyman limit.

Photometry in the Infrared

During the past 10 years, photometry in the infrared has become a very active field. Pioneer work in the field was done by Johnson and Low at the University of Arizona and by Becklin, Neugebauer, Leighton, and others at the California Institute of Technology and at the Hale Observatories. The standard blue-sensitive photomultiplier cell can be used between the far ultraviolet (0.32 micron) and the visual (0.6 micron), but it loses sensitivity rapidly as we proceed to longer wavelengths. For work in the near infrared, photocells with cathodes of the cesium oxide–silver variety work very well; they cover nicely the range from 0.6 to 1.2 microns. These near-infrared cells are operable with the same sort of electronic equipment that is used for the IP 21 cells and cooling with carbon dioxide is all that is required to eliminate excessive background noise. In the intermediate infrared (1 to 4 microns wavelength), the most effective cells to use are the lead sulfite photoconductive cells, which must preferably be cooled to the temperature of liquid nitrogen. The Earth's atmosphere absorbs quite heavily in the 1- to 4-micron range, notably near 1.8 and 2.8 microns, but there are some clear "windows" near 1.3, 2.2, and 3.4 microns. For best results one should observe in dry areas from mountain tops 9,000 feet high or more, so as to have minimum water vapor overhead.

For observations at still longer wavelengths, one goes to a completely different type of radiation receiver. Low has been very successful with a germanium bolometer operated at liquid-helium temperatures. The region between 4 and 22 microns can thus be reached, but atmospheric absorption becomes increasingly bothersome and so does irregularly distributed radiation from the night sky. The greatest blocking in this range occurs near 6.5 microns, and the relatively transparent windows used for measurement are at 5.0 and 10.2 microns. Most of the atmospheric absorption can be eliminated if observations are made from high-flying airplanes, operating at altitudes of 40,000 to 50,000 feet. The gap between optical and radio wavelengths was really bridged when Low succeeded in making some bolometric measurements at 1 millimeter wavelength!

Standards of Brightness and Color

The first major task for the photometrist—the astronomer who measures magnitudes and color indices of stars—is to establish an extensive and comprehensive network of magnitudes and colors for standard stars, distributed all over the sky and for both the Northern and the Southern Hemispheres. The establishment of such a network involves precision measurements in well-determined filter and photocell systems of constant properties. It is also required that the observer pay very careful attention to the problems of atmospheric extinction, the effects of which must be eliminated before final values of magnitudes and color indices can be arrived at. Harold L. Johnson, now at the University of Mexico, has made such measurements for the *UBV* color system, which was first suggested by him. His lists and those of his associates give data for several hundred stars, many of them at southern declinations. Supplementary lists have been prepared to include measures in the red (*R*) and in the various infrared bands (*I* and other bands). These lists are the backbone of all modern photometry. They have almost entirely replaced the earlier photographic standards, notably the famous North Polar Sequence of the early 1900's.

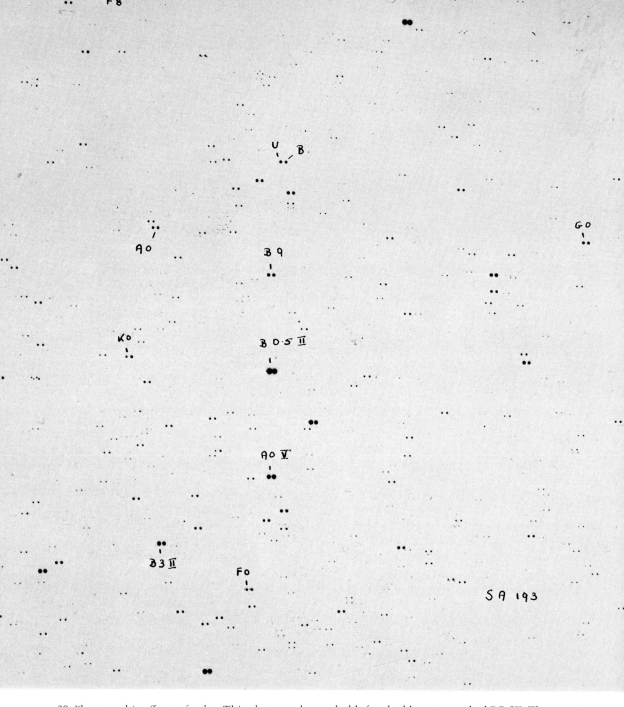

30. Photographic effects of color. This photograph shows two exposures, one, marked *U*, taken through an ultraviolet filter, the other, marked *B*, taken through the standard blue filter. To show the star images more clearly, we reproduce a negative print. The star marked K0 is a red star; its *B* image is stronger than its *U* image. The reverse holds for the blue star marked BO.5II. The remaining stars listed in order of increasing redness are marked as B3II, B9, A0, A0V, F0, and G0. A photograph made with the 40-inch reflector of Siding Spring Observatory in Australia. The southern field is that of Kapteyn Selected Area 193 (see page 65).

An astronomer who desires to obtain photoelectric substandards for any point in the sky, or who wishes to measure the UBV magnitudes for a specific star, can take as his reference star the nearest of the Johnson standards. The final result of his measurements is that, for his star, he obtains reliable values for the V magnitude and for the color indices $B - V$ and $U - B$. The next step, or steps, will depend on the nature of the research in which the astronomer is involved. When studying a variable star, he can monitor its variability by measuring at different times its V magnitude and $B - V$ and $U - B$ color indices with reference to a presumably constant standard. Or, by reference to the basic standards, he can obtain magnitudes and color indices for hundreds of stars of a special class, say white dwarfs. Since, as we noted earlier, linearity of response is built into modern photoelectric photometers, it is a simple matter to measure differences of magnitude or color index for stars differing greatly in apparent magnitude.

A different sort of problem arises when we require magnitudes and color indices for several hundred stars in a small area of the sky. This problem occurs when we study an open cluster or a globular cluster, or if we wish to measure apparent magnitudes and colors of all blue-white stars in a field along the band of the Milky Way. First we establish a *standard sequence* of magnitudes and colors involving photoelectric standards for, say, 30 stars. The sequence may well include stars in the range of visual magnitude $V = 8$ to 18, and there should be within the sequence a fair spread in colors $B - V$ and $U - B$. Next, we turn to photographic interpolation, using the photoelectric sequence as our standard reference, to find the values of V, $B - V$, and $U - B$ for all the stars in question. This is achieved by photographing the field in U, B, and V light

by the selection of suitable combinations of color filters and photographic emulsions. The photoelectric U values for the sequence stars, all of which appear on our U photograph, are used as reference standards for the measurement of the U magnitudes of all other stars for which we wish to obtain this information. Similarly, we measure the B and V magnitudes for all the stars in question. The photographic measurements can all be made in the laboratory with a photographic densitometer, an instrument that measures the size and density of the photographic images of the sequence stars. Basically we use a process of direct interpolation. It is obvious that all of these processes can be executed most efficiently by proper application of modern techniques of automation and of data processing. Figure 30 shows rather nicely some of the more conspicuous photographic effects of color. The photographic plate had two exposures, one in ultraviolet light (U), the other in blue light (B). The blue stars are readily distinguished from the intrinsically red ones.

One major weakness of the UBV and similar systems is that the filter passbands are rather wide, often too wide for detailed research on the physical properties of the stars and of the interstellar medium. In the 1950's, Strömgren and Crawford, then at Yerkes Observatory, pioneered in the establishment of suitable systems for intermediate and narrow-band photometry. The characteristics of the *uvby* system that they developed are as follows:

	Central wavelength (microns)	*Filter passband (microns)*	
u	0.35	0.038	ultraviolet
v	.41	.020	violet
b	.47	.020	blue
y	.55	.020	yellow

The *u* filter is a normal ultraviolet glass filter, but the narrow-band *v*, *b*, and *y* are specially made interference filters. Strömgren, Crawford, and Perry have established standard *uvby* values for a large number of reference stars, very much in the manner of the *UBV* standards. It is not yet feasible to make interference filters 8 × 8 inches in area, to mimic the available much smaller, 2 × 2 inches, *vby* filters and, for the present, we cannot resort to photographic interpolation to measure large numbers of stars. However, the time may not be far off when mass determination of *uvby* magnitudes and colors by photography will be possible. Already Schreur has developed a test series of gelatine filters for photographic purposes, which mimic nicely the *uvby* photoelectric system.

One disadvantage of the intermediate-band *uvby* system compared with the broad-band *UBV* system is that with a given telescope one cannot reach as faint stars in *uvby* as in *UBV*. The limiting magnitude for *uvby* is about 2½ magnitudes brighter than for *UBV*. This lack of penetration becomes even more serious when one turns to real narrow-band photometry. Here the pass bands of the interference filters often narrow down to 0.003 to 0.005 micron, and as a result another 2½ magnitudes in limiting magnitude is lost over *UBV* photometry.

A large number of secondary sequences are distributed over the sky. At several observatories in the United States, Holland, Sweden, the Soviet Union, Great Britain, and South Africa much time and effort have been spent on standard sequences in the Kapteyn Selected Areas. About 50 years ago Kapteyn of Holland selected a network of 206 centers uniformly distributed over the whole sky to provide photographic standard magnitudes. It was Kapteyn's intention that astronomers should concentrate their efforts upon obtaining colors, proper motions, radial velocities, and spectral types for the stars in these Selected Areas, in the hope of deriving from these data the characteristic properties of the Galaxy.

Why should we want to measure magnitudes and colors of thousands of stars? For the nearby stars, the color index is a direct measure of the temperature of the star. Obviously, a blue-white star with a surface temperature of 25,000°K emits a much stronger blue radiation than a red star with a surface temperature of 3,000°K. (Temperatures of stars are measured on the Kelvin scale, which has its zero at the absolute zero of temperature, 273° below 0° centigrade.) For the more distant stars, another factor comes into play. The light of such stars is appreciably reddened by cosmic dust between the stars and our Sun. This is especially so for distant stars along the band of the Milky Way. Frequently we may predict, on the basis of the appearance of the spectrum of the star (discussed at length later in this chapter), what the value of its color index must have been before the star's light was affected by the reddening from the interstellar dust; we call this the intrinsic color index of the star. If we once know this intrinsic color index, then we can obtain from the difference between the observed and the intrinsic color indices information about the amount of reddening produced by the intervening cosmic dust. We thus derive useful data regarding the absorption of light by the cosmic dust in our Milky Way system. Color-index measurements figure prominently in all studies of interstellar absorption.

Distances and Parallaxes

When we see stars in the sky that differ greatly in brightness, we are naturally curious as to how much of this can be attributed to differences in intrinsic brightness and how much of it is due to differences in distance from us. When light travels unobstructed through space, its brightness varies inversely as the square of the distance from the source. If two stars that are in reality equally luminous appear to us to differ by 5 magnitudes (which means a light ratio of 100 to 1), the fainter one must be ten times as far away as the brighter one. How can we find the distances of the stars?

The astronomer's fundamental method of finding the distance to a star is essentially the same as that of a surveyor who wants to know the width of a river. First he measures carefully as long a base line as is practicable along the bank on his side of the river. From each end of this base line he sights on a tree or other landmark on the opposite shore and measures its angular direction. This gives him the angle-side-angle triple, well known to high-school geometers, which enables him to compute any other part of the triangle. The astronomer likewise must first carefully measure his base line and he must then measure with high precision the direction angles at which he sees the star at opposite ends of his base line. We have not enough room on the Earth to provide us with a proper base line, one long enough for the measurement of the very great distances to the stars. But the Earth moves in a very large ellipse (almost a circle) around the Sun (Fig. 31). If we but wait 6 months after one pointing, we shall have moved, without any effort on our part, 186,000,000 miles (299,000,000 kilometers), or twice the distance from the Earth to the Sun. From the two ends of this base line, the astronomer sights on a number of stars and he will find that at least the nearest ones will have shifted their positions measurably with reference to the base line, or, more precisely stated, with reference to the background of much more distant stars.

The astronomer uses as indicators of distance the shifts in the apparent positions of the stars. By definition the *astronomical unit*,

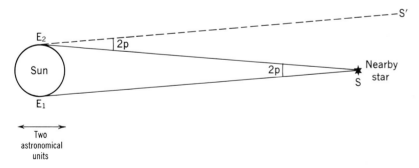

31. The parallax of a star. When the observer on the Earth moves in half a year from position E_1 on the Earth's orbit to position E_2, a distance apart equal to 2 astronomical units, the nearby star will have apparently shifted its position among the more distant stars, from direction E_2S' (parallel to E_1S) to E_2S. This angle ($2p$) is equal to the angle at the star subtended by twice the astronomical unit or is then equal to twice the star's parallax.

the mean distance from the Earth to the Sun, which is 93 million miles or 149.6 million kilometers (1 mile = 1.609 kilometers), is taken as the base line, and the angular shift that corresponds to it at the distance of the star is called the *parallax* of the star. Hence, the parallax can be defined as the angle subtended by the mean distance from the Earth to the Sun, as viewed from the star. We note that the total displacement shown by a star is equal to twice its parallax, since the distance apart of the two extreme positions of the Earth for the measurement of the parallax effect equals 2 astronomical units.

It is not difficult to compute the distance of a star from its parallax. To do so, we introduce a new unit of distance, to which has been given the hybrid name of *parsec*. A star at the distance of 1 parsec has a *parallax* of 1 *second* of arc. The distance in parsecs is equal to the reciprocal of the parallax in seconds of arc: $d = 1/p$. The distance of 1 parsec is equal to 206,265 astronomical units, or $206,265 \times 93,000,000 = 19,200,000,000,000$ miles. Instead of the parsec, the more picturesque unit, the *light-year*, is often used. The light-year is the distance that light travels in 1 year. At the rate of 186,000 miles per second, and with about 31,600,000 seconds in a year, the light-year is equal to 5,880,000,000,000 miles. One parsec is, therefore, equal to approximately 3.26 light-years. The nearest star has a parallax of 0.75 second of arc, hence a distance of 4.30 light-years; light takes $4\frac{1}{3}$ years to reach us from our nearest neighbor. We can also say that this star is at a distance from us of $1\frac{1}{3}$ parsecs.

Finding the parallax of even one star is an exacting and time-consuming process. Photographs must be taken with a long-focal-length photographic telescope in order to have a large scale on the photographic plate. These photographs should be repeated at several 6-month intervals to separate the effect of the star's own motion, which is a uniform motion along a straight line on the photographic plate, from the parallax effect, which goes through its cycle once every year. If the star for which the parallax is being sought is much brighter than the stars near it, then we must develop some method of cutting down its light, since it is impossible to measure accurately relative positions of photographic images that are of very different sizes and densities.

The modern photographic method for the measurement of stellar parallaxes was developed about 70 years ago by Schlesinger, with the aid of the 40-inch Yerkes Observatory refractor. Measured parallaxes are available for some few thousands out of the many millions of stars; modern parallax catalogs list parallaxes of reasonable reliability for about 8,000 stars, and additional trigonometric parallaxes are presently being produced at a rate of only 60 to 80 per year. With the best of care, the astronomer cannot measure the parallax of a star with an error of less than 0.004 second of arc, and larger errors are not uncommon. Hence, the measured trigonometric parallax for a star that is only 50 parsecs distant may have a value anywhere in the range 0.016 to 0.024 second of arc, and the corresponding percentage uncertainty in the estimated distance is quite large. There is some hope for improvement in precision of the basic measurements. Van Altena at Yerkes Observatory is developing new techniques of measurement and reduction that he hopes will reduce the errors by 50 percent. High-quality trigonometric parallaxes are obtainable only for relatively nearby stars, and in practice parallax observers wisely concentrate their efforts upon stars within 20 parsecs of the Sun. Fortunately, there is quite a variety of stars within this distance.

In the past many observatories took part in the measurement of trigonometric parallaxes. Among the early leaders in the field, most of which are still active in this area, are the Yerkes, Allegheny, Leander McCormick, Sproul, and van Vleck Observatories in the United States, the Royal Greenwich Observatory, and the Royal Observatory at the Cape of Good Hope. All of these have traditionally concentrated their efforts on stars brighter than the tenth or, at the faintest, the twelfth magnitude. Several newcomers can now be added to the list. The most famous telescope now active in parallax work is the 61-inch reflector of the U.S. Naval Observatory at Flagstaff, Arizona. This marvelous precision astrometric reflector was constructed principally for the determination of trigonometric parallaxes of stars between the twelfth and the seventeenth magnitudes. The telescope has an annual yield of 40 precision trigonometric parallaxes. The Lick 36-inch refractor and the large refractor at Pulkova Observatory in the U.S.S.R. are also engaged upon parallax work. Luyten has attempted some studies in the field for very faint stars, mostly blue ones, with the Palomar Schmidt telescope.

An observer of trigonometric parallaxes often wishes that he might inhabit Jupiter, whose orbit has a diameter equal to 10 astronomical units, or Pluto, which would give an orbital base line of 75 astronomical units. The long-range future looks bright for the parallax observer, for it should not be many years before earth satellites will be able to bring back—or telemeter back—from outer space data on parallax shifts from stations at distances of 5 or more astronomical units from the Sun! This undertaking represents one of the most important space missions of the future.

All parallaxes determined photographically are measured relative to the average for the background stars on a photographic plate with the parallax star near its center. A small systematic correction is generally applied to correct the relative parallax to a true one. The amount of this correction is, however, uncertain. In the future, it might be well to check this correction with great care. Attempts should be made to use one or more faint galaxies as basic reference points for zero parallax. Galaxies are millions of parsecs away and hence certainly show no measurable parallax displacements of their own.

But before we get too far out into space, let us first see what results can be found from direct parallax measurements.

Absolute Magnitudes

If we know the distance of a star and also how bright it appears to us, we can find its true or intrinsic brightness. We may then compare the intrinsic brightness of each star with that of our Sun as the standard and obtain what is called the *luminosity* of the star. Or we may imagine all stars placed at the same distance and compute, from the observed apparent magnitude and the known actual distance, what the magnitude of the star would be if it were placed at this standard distance. We call this the *absolute magnitude* of the star. Long ago international amenities led us to choose 10 parsecs as the standard distance. The formula for the computation of the absolute magnitude is

$$M = m + 5 - 5 \log d,$$

where m is the apparent magnitude and d is

the distance in parsecs.* Altair, with a measured parallax of 0″20, has an apparent magnitude $m = 0.9$. Since its distance is $1/0″20 = 5$ parsecs, its absolute magnitude is

$$M = 0.9 + 5 - 5(0.70) = +2.4.$$

When, hypothetically, we place all the stars at the standard distance of 10 parsecs, we find that they show as wide a range in absolute magnitude as they do in apparent magnitude. The values of the observed absolute magnitudes range all the way from $+18$ to -10. In our imagination, we can set the Sun at the standard distance of 10 parsecs and we find that its absolute magnitude lies near the middle of the range, at $+4.7$; our Sun would appear as an inconspicuous star of very nearly fifth apparent magnitude if it were placed at the standard distance of 10 parsecs. If we compare any other star with the Sun as the standard—the standard candlepower, as it were—we obtain what is called the *relative luminosity* of the star. Some of the brightest stars are 14 to 15 magnitudes brighter intrinsically than our Sun; they are pouring forth $2.512^{15} = 1$ million times as much energy as the Sun. On the other hand, the faintest known star in absolute magnitude is about 14 magnitudes fainter than the Sun; the star is emitting $1/300,000$ of the radiation emitted by the Sun. The ratio of intrinsic brightness between the brightest and the faintest stars is

therefore at least of the order of 100 billion! This range is all the more remarkable since the corresponding ratio of masses is less than 10,000, probably not much over 1,000. There are such tremendous differences from one star to another that it is unsafe to guess whether a particular star is faint because of its low intrinsic luminosity or because it is very far away. On the average, the fainter stars will be the more distant ones, but we cannot say anything about the distance of a particular star unless we know its absolute magnitude.

Spectral Classification

Spectroscopic tools are essential to the astronomer. All large telescopes are fitted with spectrographs of various designs, and large prisms that can be placed in front of the objectives of photographic telescopes (Fig. 32) are standard equipment at several observatories. Figure 33 shows the kind of photograph that is obtained when one of these objective prisms is attached. By a slight trailing of the star's image in the direction at right angles to the dispersion, the stellar spectra, with few exceptions, are made to appear as little bands, crossed by dark lines, the so-called absorption lines. Even the most cursory inspection shows that not all spectra are alike. Certain lines of hydrogen, called the Balmer lines, are strong in some spectra, weak in others, and entirely absent from the spectra of still other stars. The lines of iron and other metals are sometimes present, and in some spectra molecular bands are the outstanding feature. These differences were a challenge to the astronomers of the late nineteenth century, when the various spectral characteristics were beginning to be studied. It was soon clear that the stars could be subdivided into a fairly small number of spectral classes, which merge gradually one into the next.

---

*To derive this equation will be easy for those who are used to logarithms and remember how we define magnitudes. Let l be the apparent brightness of the star and L its absolute brightness. Then

$$\frac{l}{L} = \frac{(\text{standard distance})^2}{(\text{actual distance})^2} = \frac{10^2}{d^2} = 2.512^{M-m}.$$

Taking logarithms, $0.4(M - m) = 2 - 2 \log d$, and hence

$$M = m + 5 - 5 \log d.$$

32. Mounting an objective prism. Dr. Freeman D. Miller, mounting the 6° objective prism (showing his reflection) on the corrector plate of the Curtis-Schmidt telescope.

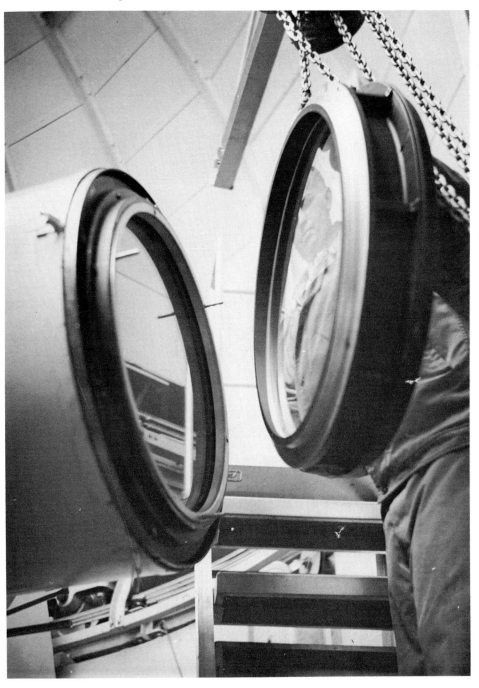

33. Examples of spectral classification. Representative spectrum-luminosity classes are indicated for 10 stars in a southern field at right ascension 8^h24^m, declination 18°41′ south (1950). The telescope is the Curtis-Schmidt telescope of the University of Michigan (Fig. 32) now mounted at the Cerro Tololo Inter-American Observatory. The plate was selected by Dr. D. J. MacConnell, Department of Astronomy, the University of Michigan, who made the following classifications: 1, K3III; 2, F2IV; 3, A1IV; 4, A9IV; 5, K0III; 6, B2IV; 7, M2III; 8, B9IV; 9, F6V; 10, G2V. (University of Michigan Observatory photograph.)

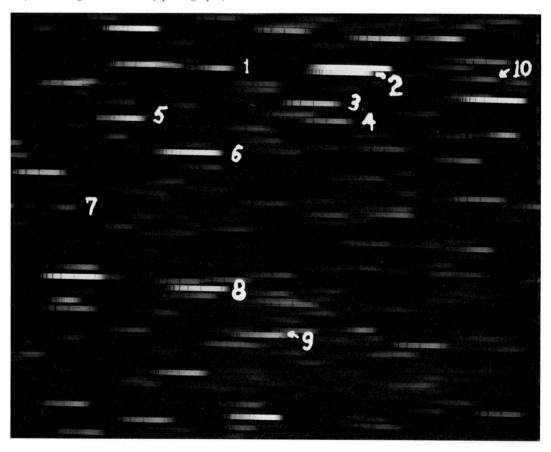

The classification that has been widely used for a long time was worked out at the Harvard Observatory under Pickering by Miss Cannon, Miss Maury, and Mrs. Fleming. The *Henry Draper Catalogue* is the work of Miss Cannon; it contains the spectral classifications of 225,320 stars of both the northern and the southern sky, and includes practically all the stars to a limit between the eighth and ninth apparent magnitudes. The *Henry Draper Extension*, which was completed after Miss Cannon's death by Mrs. Mayall, continues the classifications to the eleventh magnitude for special regions of the sky. The spectral classes were first lettered in alphabetical order, but by a "survival-of-the-fittest" development the sequence of classes has narrowed down to O, B, A, F, G, K, M. A few odd stars of classes W, R, N, C, and S, and some individuals craving for notoriety that are labeled "pec" for peculiar, supplement the main series. The O to M sequence is very strikingly a line-up in color: the O and B stars, such as Rigel and other stars in Orion, are blue; the A stars, like Sirius and Deneb, are white; Procyon and Capella, of the F and G types, are yellow; finally, the orange and red K and M stars have Arcturus, Aldebaran, Antares, and Betelgeuse as shining examples.

The surface temperature of a star, or rather the degree of ionization in its atmosphere, is the determining factor that decides where a star shall be placed in the spectrum line-up of the Henry Draper Classification. Miss Cannon found it necessary to subdivide the B stars on a decimal system, with the hottest listed as B0, B1, B2, . . . and the coolest as B8 and B9, and with B9 not very different from A0. Miss Cannon did not distinguish in her main catalogs between an intrinsically very luminous B0 star, a normal B0, and a subluminous B0 star, but the notes at the end of her catalogs show that she was aware of the presence of subtle differences indicative of absolute-magnitude effects. Miss Maury's classification, cumbersome and now wholly abandoned, was in reality a two-way classification system in which luminosity effects figure along with the temperature, or color, sequence. Chemical composition is another complicating factor. Differences in chemical composition from one stellar atmosphere to the next produce slight differences in spectral appearance. Subnormal strength of metallic lines for a given spectral class was noted in the first system of spectral classification as indicative of unusual chemical composition of a star's atmosphere. For the moment we shall leave out of consideration the effects of luminosity and chemical composition and concentrate on the main stream of spectral classes, which is a temperature, or color, sequence.

The O stars are the hottest, some with temperatures as high as 100,000°K. Their spectra can be easily recognized by certain characteristic lines emitted by ionized atoms (atoms that have lost one or more of their outer electrons) and by the far-violet extension of the spectra. Similar in spectra and temperature to the O stars, but with bright lines—emission lines—in their spectra, are the group called the Wolf-Rayet (W) stars. The emission lines are produced mostly in extended and expanding outer shells. The B stars, blue in color and with surface temperatures of the order of 15,000° to 30,000°K, show dark absorption lines of helium and hydrogen. Some B stars also show emission lines. Helium fades and hydrogen strengthens as we approach class A along the spectrum series. Lines of calcium and other metals, such as iron and magnesium, gradually increase in strength through classes F and G. Our Sun is a typical G star. In class K the calcium lines become very strong, and bands due to molecular compounds come into view. Class M stars are red, with temperatures of less than 3,000°K; their spectra show absorption bands of titanium oxide. The symbols R, N, and S refer to a series of cool stars (parallel to the K–M series) in whose spectra other compounds are present.

There is a wide range in absolute magnitude for stars of most spectral types. Capella, for example, has a spectrum very much like that of the Sun, but its absolute magnitude is +0.4, which means that Capella is 4 to 5 magnitudes (or close to 100 times) brighter intrinsically than the Sun. The surface temperatures of the two bodies are nearly the same, but their physical conditions are quite different. A *giant star* such as Capella has a much more extensive atmosphere, of much lower density, than a *dwarf star* like the Sun. These differences of density and pressure cause some spectral lines to be stronger in the giants and others to be enhanced in the dwarfs. Adams and Kohlschütter at the Mount Wilson Observatory in 1914 were the first to study the spectra of known giants and dwarfs of the same spectral class. Once the luminosity effects were noted it was possible to assign all stars of that same spectral type to either the giant or the dwarf branch, especially in the case of stars of spectral classes F to M. Adams and Kohlschütter's work was preceded by that of Miss Maury, who really discovered the first criterion of absolute magnitude. She found that there were some stars with unusually sharp spectral lines, which she designated by c; a star classified as cB3, for example, is a sharp-line B3 star. Hertzsprung showed that this characteristic is indicative of high intrinsic brightness.

In the Henry Draper system of classification, Miss Cannon introduced the practice of remarking on the sharp-line c stars, thereby indicating the most notable supergiants. For the dispersion used in the Henry Draper Classification it did not seem feasible to do more. However, Bertil Lindblad and his associates at Stockholm and Uppsala found that there were certain features in the spectra of very low dispersion for faint stars that are indicative of the star's luminosity. They developed techniques for distinguishing between giants and dwarfs among the stars of a given spectral type. In the spectral catalogs prepared after the *Henry Draper Catalogue*, notably those of the 1920's and '30's published by the Hamburg-Bergedorff, Potsdam, and McCormick Observatories, giant-dwarf separation was generally made. Yet even today the *Henry Draper Catalogue* stands first; its greatest value lies in its homogeneity and completeness.

The modern trend in spectral classification is to use an objective prism with a fairly large angle and thus produce stellar spectra in which considerable fine detail can be observed (Fig. 33). By the use of higher dispersion, we obtain greater resolution in the spectra, but it is not possible to reach very faint stars, and we are troubled by overlapping of spectra in dense regions of the Milky Way. The increased accuracy of classification, however, offsets for many problems the loss in penetrating power.

At the Yerkes Observatory, Morgan and Keenan, assisted by Kellman, developed a two-dimensional system of classification. Morgan studied the whole range of spectral types with the idea of finding "natural groups," the purpose being to use such groups in probing the galactic system. These groups are characterized by a narrow range in luminosity and by easily identified criteria in spectra of low dispersion. The latest version of the

Atlas of Stellar Spectra (by Morgan and Keenan; known to astronomers as the MK *Atlas*) is of great value for all work on accurate classification. Six luminosity classes are recognized (Fig. 36): Ia, most luminous supergiants; Ib, less luminous supergiants; II, bright giants; III, normal giants; IV, subgiants; V, dwarfs.

We shall see in later chapters that the MK *Atlas* and its system of spectrum-luminosity classification have proved invaluable for the study of our Milky Way system. They have formed the basis for the spectrum–absolute-magnitude classifications carried out by Morgan in collaboration with the Warner and Swasey Observatory. Our modern views on the spiral structure of our Milky Way system are largely based upon studies involving extensive applications of the MK system. We are not yet certain of the precise absolute magnitudes that should be assigned to each luminosity class, especially for the absolutely brightest stars, but already the success of the MK system has been so great that it supplants practically all earlier work.

In recent years astronomers have become so deeply involved with problems of spectral classification that several independent efforts have been made to improve upon the basic Henry Draper System. In these new approaches to spectrum-luminosity classification total intensities of certain spectral lines—the Balmer lines for instance—and brightness distribution in the continuous background are used as classification parameters. These can be defined numerically if we use the techniques of intermediate- and narrow-band photoelectric photometry. For example, the Strömgren, Crawford, Graham technique measures the strength of one of the Balmer lines of neutral hydrogen, H-beta, by the use of two filters, one with a 0.003-micron-wide

passband, the other with a passband 0.015 micron wide. The first measures only the total intensity of the H-beta absorption line, whereas the second measures in addition quite a bit of the relatively clear continuous background. The difference of the two brightnesses appears to be dependent mostly on the absolute magnitude of the star, at least for B and A stars. The Chalonge criterion of the drop in intensity at the Balmer limit—near 3650 angstroms, where many Balmer lines crowd together—is measured in the Strömgren *uvby* system of intermediate-band photoelectric photometry. The *u* filter is centered close to the Balmer limit, the *v* filter at somewhat greater wavelengths outside the region of the crowding of the Balmer lines. The $u - v$ color index measures photoelectrically the total strength of the crowded Balmer lines.

In modern approaches to spectral classification, increasing emphasis is being placed on the inclusion of criteria in the ultraviolet. The MK *Atlas* is being supplemented by the recent (1968) *Atlas of Low-Dispersion Grating Spectra* by Abt (Kitt Peak Observatory), Meinel (University of Arizona), and Morgan and Tapscott (Yerkes Observatory). The spectra, obtained with a grating spectrograph capable of recording spectral lines in the ultraviolet almost to the Balmer limit, provide for quite a few more criteria than were attainable with the MK *Atlas*. Figures 34 and 37 show samples of spectra from the 1968 *Atlas*. Precision spectrum-luminosity classification is done by the fitting of the spectrum of a given star into the general sequence of the *Atlas*.

In the past 20 years spectral classification in the infrared has made rapid strides. Nassau, McCuskey, Blanco, and their associates at the Warner and Swasey Observatory have been the most active workers in the field in the Northern Hemisphere, whereas Henize,

34. Spectra of main-sequence stars with classes B0 to M5. The four photographs are from Abt, Meinel, Morgan, and Tapscott, *Atlas of Low-Dispersion Grating Stellar Spectra* (Kitt Peak National Observatory, 1968). These are negative prints with the normally dark absorption lines shown white in the photograph; this makes the spectral features stand out better than on normal prints. The caption for each photograph is that shown in the *Atlas*.

Main Sequence
O9 Ⅴ – A0 Ⅴ

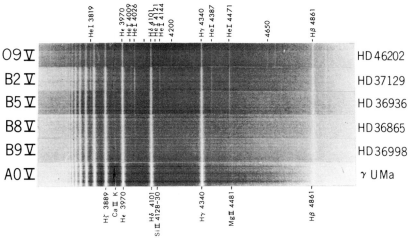

The hydrogen lines increase in intensity with spectral type and reach their maximum strength near A0 Ⅴ. The lines of He I have a sharp maximum at B2 Ⅴ and decrease smoothly in intensity thereafter. The stellar K-line is first used for classification near B8 and increases rapidly in intensity toward later types.

Main Sequence
A0 Ⅴ – F5 Ⅴ

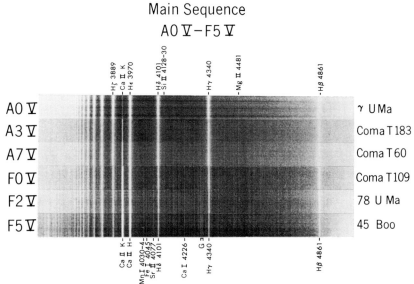

The ratio of Ca Ⅱ K/Hδ changes rapidly with spectral type, and the neutral metallic lines grow stronger. The G-band first appears as a continuous feature near F2.

Main Sequence
F5 V – K2 V

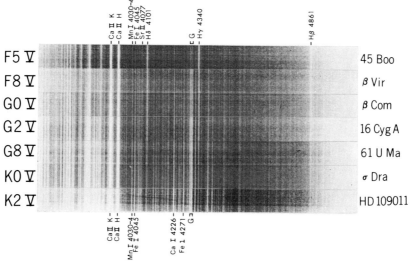

F5 V	45 Boo
F8 V	β Vir
G0 V	β Com
G2 V	16 Cyg A
G8 V	61 U Ma
K0 V	σ Dra
K2 V	HD 109011

The hydrogen lines decrease in intensity, while the neutral metallic lines increase.
The G-band grows in intensity to late G-types; its appearance alters thereafter.

Main Sequence
K2 V – M5 V

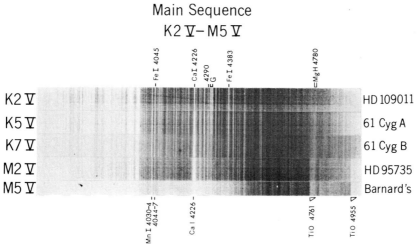

K2 V	HD 109011
K5 V	61 Cyg A
K7 V	61 Cyg B
M2 V	HD 95735
M5 V	Barnard's

The G-band changes in appearance, and the line Ca I 4226 increases rapidly in
intensity with advancing type. The bands of Ti O appear near M0 and grow with
decreasing temperature. An absorption band of MgH, centered near 4780 is well marked
in late K-type dwarfs; it is blended with the TiO absorption near 4761 in the M dwarfs.
The MgH band is an important luminosity discriminant.

35. Luminosity effects at B1. Negative prints are shown. (Reproduced from Morgan and Keenan, *Atlas of Stellar Spectra*, Yerkes Observatory, University of Chicago.)

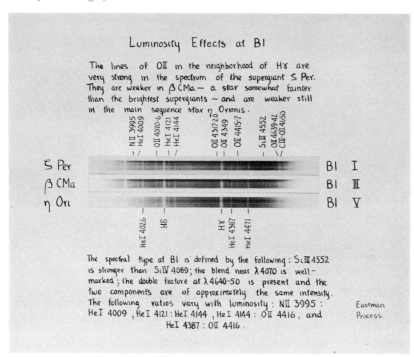

36. Luminosity classes. In the classification system
first developed by Morgan, Keenan, and Kellman,
one distinguishes both spectral type and luminosity class. The schematic diagram gives vertically
the approximate visual absolute magnitudes for
the stars of different spectral types and luminosity classes (indicated by roman numerals). (From
an unpublished compilation by Matthews.)

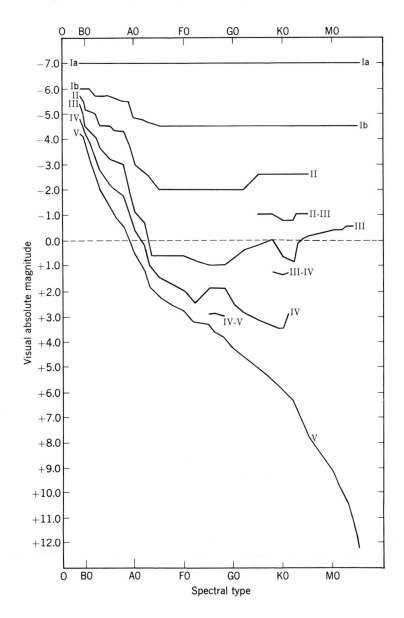

37. Spectra of hot supergiant stars. These representative spectra from the *Atlas* by Abt *et al.* show beautifully the very sharp and narrow absorption lines observed in the most luminous supergiant stars; negative prints.

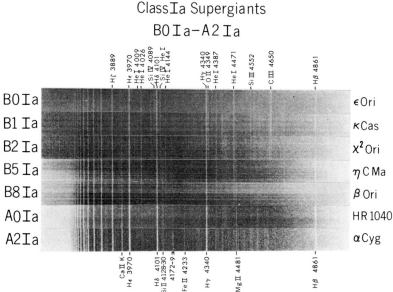

The hydrogen lines are weaker than in other luminosity classes. The lines He I 4471 and 4026 have an extremely flat maximum. Lines of O II are at maximum strength at Class B1. Lines of Fe II become outstanding in Class A.

Haro, and Westerlund have classified extensively for the Northern and Southern Hemispheres. The carbon stars and the S stars are, besides the M stars, the most common varieties of red stars. In the near infrared, Keenan distinguishes the late M stars by the presence of titanium oxide and vanadium oxide molecular bands and the S stars by zirconium oxide bands. The carbon stars—now generally referred to by the symbol C—were previously called the R and N stars. The C stars have characteristic bands of the carbon molecule and of the cyanogen (CN) molecule.

Proper Motions

In 1718, Halley compared his observations of the positions of Arcturus and Sirius with the positions as given in the catalog of Ptolemy, and he found that these stars had moved by appreciable amounts. The so-called "fixed" stars are not stationary but are constantly changing their positions in the sky. Figure 38 illustrates the motion of the star with the largest known annual displacement. This star, called Barnard's star after its discoverer, moves at the rate of 10 seconds of arc

38. A fast-moving star. Two photographs of Barnard's star, taken 11 months apart with the 24-
inch refractor of the Sproul Observatory. In making the combined print, the second plate was
shifted slightly with respect to the first. The arrow
points to one pair of images for which the displacement differs from that for the others; we deduce that this star has changed its position per-
ceptibly during the interval of 11 months.

per year, a Moon's diameter in less than 2 centuries. When two photographs taken a year or
so apart are combined, as in Fig. 38, the runaway can be spotted.

Most stars move very slowly across the sky,
so that long intervals of time are necessary for
the detection of their motions. Accurate observations of position in the sky are made
with the meridian circle; from the altitude at
which the star crosses the meridian at a given
observatory we deduce its declination, and
from the precise timing of the meridian passage we obtain its right ascension. The annual
proper motions of a star in declination and in
right ascension are obtained by dividing the
total displacements in declination and in
right ascension by the interval in years that
has elapsed between the measurements of the
two positions. Proper motions are commonly

measured in seconds of arc per year. The total
displacement increases proportionally with
time, so that, if 10 or 20 years do not suffice to
show a measurable quantity, 50 or 100 years
may do the trick.

The effective limit for visual observation
with meridian circles lies at about the ninth
apparent magnitude. To measure the proper
motions of the fainter stars, we turn to photography. Suppose we have two photographs of
the same part of the sky taken with the same
telescope some 20 or 40 years apart. If the
scale of the plate is large enough, we can
measure the relative positions of the stars to
within 0.01 second of arc. The differences between pairs of positions can be reduced to
proper motions if there are enough stars on
the plate with known proper motions determined by meridian-circle observations.

For all the bright stars and many of the faint ones the proper motions are now accurately known. For many years the catalogs prepared under the direction of Lewis and Benjamin Boss at the Dudley Observatory in Albany and by H. R. Morgan at the U.S. Naval Observatory figured prominently in researches on positions and proper motions, but these catalogs are now being replaced by others of greater precision. The Smithsonian Astrophysical Observatory has published a comprehensive *Catalog* with proper motions for nearly 260,000 stars. Photographically determined proper motions have been obtained at the Royal Observatory at the Cape of Good Hope, at the Yale and the McCormick Observatories, and at various German observatories under the auspices of the Astronomische Gesellschaft. One of the most recent accomplishments has been the undertaking by the German observatories under the leadership of Heckmann, Fricke, and Dieckvoss to determine anew the positions of the stars in the catalogs of the Astronomische Gesellschaft and we now have in the *Catalog AGK3* proper motions for 180,000 stars, mostly between eighth and twelfth magnitudes; average accidental errors amount to ±0.008 second of arc.

In the preparation of reliable catalogs of proper motions, one of the principal concerns of the astronomer is to find a reference system that is as nearly fixed in space as possible. All measured proper motions are relative to some reference frame, be it of faint and therefore presumably distant stars, or of asteroids, the motions of which are predictable from dynamical theory. There are, however, many complicating factors. The rotation of our Milky Way around the center in Sagittarius produces, for example, systematic motions for even the most distant stars of our Milky Way system.

In recent years there have been many advances toward the establishment of an improved fundamental reference system of stellar positions and of proper motions. According to Fricke of Heidelberg, who is one of the leaders in the field, the best published proper motions for eleventh-magnitude stars now have accidental errors of the order of ±0.008 second of arc and systematically they should be correct to within ±0.004 second of arc. Within a decade it should be possible to halve both errors and from then on we should be progressing rapidly on the road to perfection.

In the 1940's research was initiated toward the measurement of proper motions with reference to objects entirely outside our galactic system, namely, faint galaxies. The Lick Observatory Survey was begun by Wright in the late 1940's. The first-epoch plates were taken with the Lick 20-inch Ross astrograph in the early 1950's, mostly by Shane and Wirtanen, and the first second-epoch plates are being taken 20 to 25 years later. The measurement of the proper motions of the faint stars, mostly of fifteenth to seventeenth magnitudes, is being done by Vasilevskis and Klemola. Since the minimum distances of the reference galaxies are of the order of 100 million parsecs, the crosswise motions of these galaxies can under no circumstances amount to as much as 0.00001 second of arc per year, and the reference frame of the faint galaxies is therefore immovable and stable to a much closer tolerance than the precision of 0.001 second of arc per year that is the sort of stability we ask for in our measurements of faint stars. An independent parallel survey that is about to yield results is that undertaken by Deutsch and associates at Pulkova Observatory in the U.S.S.R. In the Pulkova Survey photographic plates are centered on bright galaxies with distances in the range 2 to 100 million parsecs. The reference points of un-questioned stability are the knots in the spiral and elliptical galaxies. As is often the case, the Southern Hemisphere has initially received much less attention than the Northern. However, Yale and Columbia Observatories have established a station in Argentina with a telescope nearly identical with the Lick 20-inch Ross Astrograph. First-epoch plates are now being taken, and in 25 years we may expect to have available many new precision proper motions for southern declinations.

In the early 1970's a new approach was developed for the measurement of fundamental star positions and proper motions. Radio astronomers perfected and applied a technique for measuring positions of radio sources with very high precision. By the use of interconnected arrays of three or more steerable radio telescopes, researchers measure the declinations of radio sources with a precision of the order of $\pm$ 0.01 second of arc and without any reference to optical standards. Differences in right ascensions between radio sources can also be measured with comparable precision, even though the radio sources may be far apart in the sky. Since they know the right ascensions of several radio sources identified with the optical equivalents of these sources, investigators can derive a reference zero-point for the right ascensions of all radio sources. There are available already radio and optical positions for about 100 radio sources, which possess optical equivalents and which are either radio stars (Algol and Beta Lyrae are radio stars!) or radio galaxies that have been identified optically. The fundamental reference system of star positions and proper motions for the future will almost certainly represent a system based in part on optical and in part on radio data.

For a star with a known distance of d parsecs, it is possible to translate the angular value of the proper motion in seconds of arc per year, which is indicated by the Greek letter μ (mu), into a linear velocity expressed in kilometers per second, V. The simple formula reads:

$$V = 4.74\mu d.$$

A star with an annual proper motion $\mu = 0.1$ second of arc per year, at a distance $d = 50$ parsecs from the Sun, will have a linear cross-wise velocity $V = 23.7$ kilometers per second. The linear velocities of the stars range generally from a few to 100 kilometers per second. Higher values are rarely found and the average linear velocities are about 20 kilometers per second. Since observations show that the average crosswise linear velocities are about the same for nearby and for distant stars, the proper motions offer a fairly good distance criterion. The nearby stars have, on the average, considerably larger proper motions (in seconds of arc per year) than the more remote ones. We can be fairly certain that a star which stands out from its neighbors by its large proper motion has a good chance of being relatively nearby, and this star should be a good candidate for having its name placed on a program for the measurement of trigonometric parallaxes. The program for the U. S. Naval Observatory 61-inch reflector, for example, is almost entirely limited to stars with observed proper motions of 0.2 second of arc per year or greater.

To select the faster-moving stars for subsequent and accurate parallax measurements, the astronomer uses a device known as a "blink microscope." Two plates of the same region taken some years apart are so arranged that first one and then the other is viewed through the eyepiece of the microscope. When the alternation takes place quickly enough, the main pattern of stars appears unchanged but the stars that have a large angular motion will apparently hop back and forth in the field. We can thus distinguish those stars that are likely to be our nearest neighbors from the general run of stars. In Chapter 3 we shall see how useful it is to select these nearest neighbors and we shall discover how crowded—or rather, just how empty—is our particular part of space.

Radial Velocities

When we deal with proper motions, we are concerned only with the angular displacements of the stars projected upon the celestial sphere. Proper motions tell us nothing about the velocities of the stars in the line of sight. If the stars are moving in space, some must be getting closer to us, others farther away. To measure these line-of-sight velocities, or *radial velocities*, we turn again to the spectrum, which yields radial velocities measured in kilometers per second.

Since light is a wave motion, it is not surprising that it has some of the characteristics of sound. We have all noticed how the pitch of an automobile horn drops suddenly as an approaching car passes. If a star is moving toward us, the wavelength of the light we receive from it is shortened and the absorption lines in its spectrum shift their positions toward the violet end of the spectrum. When the star is moving away from us, the lines are shifted toward the red. The shift in wavelength is proportional to the relative speed of the star and the observer. This correlation, known as the Doppler effect, is expressed by the simple equation

$$\frac{\lambda - \lambda_0}{\lambda_0} = \frac{v}{c},$$

where λ is the observed wavelength of a line

in the spectrum of the star, λ_0 the laboratory standard rest wavelength of the same line, v the line-of-sight or *radial velocity* of the star, and c the velocity of light (186,000 miles per second).

How do we measure the shift in wavelength for a given star? Figure 39, which is a reproduction of two spectra of Castor, shows the stellar spectra with a laboratory standard spectrum photographed on either side. The displacement resulting from radial velocity can readily be seen by direct inspection of the photograph. The spectrum-line shifts are affected by the motion of the Earth as well as by that of the star itself. The effects due to the Earth's rotation and revolution can, of course, be calculated and removed. The radial velocities of stars listed in catalogues have been corrected for the Earth's motion; they are listed as "reduced to the Sun."

One complication that sometimes enters into the measurement of radial velocities arises from the fact that many of the stars are found to have velocities that change in a regular cycle over a definite period of time. A companion star—often invisible—may be present and the two stars revolve about each other under their mutual gravitation. These spectroscopic double or binary stars are very interesting in themselves and from them we derive information about the masses of the stars, but they have delayed considerably the determination of the radial velocities of many fainter stars. Enough spectral photographs of a star with variable radial velocity must be taken to find out how much of the observed speed is due to the orbital motions in the system and how much to the velocity of the system as a whole. The two spectra of Castor shown in Figure 39 prove that the component of Castor for which the spectra are shown is itself a spectroscopic binary.

39. Radial-velocity effects in Castor. The bright bands in the center crossed by dark lines are two spectra of the same component of Castor taken at different times. The bright lines on either side are the comparison spectrum, photographed at the same time as the star's spectrum to serve as a standard of reference. Each component of the double star Castor is itself a spectroscopic binary. The upper one of the star's spectra shows a radial displacement due to a velocity of recession of 24 miles per second; the lower one shows a radial-velocity displacement of approach amounting to 32 miles per second. (Lick Observatory photograph.)

Much of the early radial-velocity work was done by Campbell, Wright, and Moore at the Lick Observatory with the 36-inch refractor and the three-prism Mills spectrograph. The Lick *Catalogue of Radial Velocities* includes almost all naked-eye stars in both hemispheres. Several observatories are now engaged upon the measurement of radial velocities of the fainter stars. We require great numbers of radial velocities for the detailed study of the Galaxy, and the task of obtaining them has been a vast one. In addition to the leading American observatories, Canadian, French, and Soviet observatories have been most active in the gathering of data for the Northern Hemisphere; the Dominion Astrophysical Observatory in British Columbia deserves honorable mention. In the Southern Hemisphere, radial-velocity measurements have been made especially in South Africa at the Cape and Radcliffe Observatories, and at Mount Stromlo Observatory in Australia. From time to time, radial-velocity catalogs have been prepared, which bring together all available data. In 1953, R. E. Wilson of Mount Wilson and Palomar Observatories (now Hale Observatories) issued such a *General Catalog*, with entries for 15,107 stars. A new catalog has been assembled under the direction of Evans of the University of Texas and another one is in preparation at Kitt Peak National Observatory by Abt.

Modern instrumental developments are helping to speed up the acquisition of data. Semiautomatic techniques for the measurement of the spectral photographs, which employ modern computer methods of data processing, are making it possible to reach faint stars and measure their radial velocities with high precision. For the brighter stars of solar type, we can now obtain radial velocities with accidental and systematic errors of the order of ± 1 kilometer per second and less. The precision goes down by a factor 2 or 3 when we go to fainter stars or measure the radial velocities of stars with few sharp lines in their spectra, such as the A and B stars, but no self-respecting observatory involved in radial-velocity work likes to publish radial velocities with probable errors in excess of ± 5 kilometers per second.

Every stellar spectrograph has four basic components. At the focus of the telescope, the image of a star is set upon a narrow *slit*. A *collimator*, of lens or mirror type, is used to focus the illuminated slit upon the *dispersive element*, now generally a precision *line grating*, and the resulting spectral band is photographed with the aid of a *camera*, now not infrequently a fast Schmidt-type camera. By placing the telescope at a mountain site with excellent seeing conditions, one obtains small star images and the slit can be closed down to a narrow width and still transmit most of the star's light. The grating spreads the light into a spectrum with minimum light loss and the high-quality camera records it all with precision. Mirrors are now generally covered with coatings of aluminum instead of silver, which assures permanent high reflectivity even in the ultraviolet; the camera photographing the spectrum has either highly reflective mirrors or lenses made of transparent optical glass, covered with a transparent antireflection coating. The photographic manufacturers, notably the Eastman Kodak Company, have traditionally provided astronomers with photographic emulsions of great sensitivity, capable of recording the spectra of the stars for each desired wavelength interval. At Kitt Peak National Observatory, the fast ultraviolet spectrograph on the 36-inch reflector is capable of recording the spectrum of a seventh-magnitude star with a *dispersion* of 63 angstrom units per millimeter inside 6 minutes; an eleventh-magnitude star would yield a spectrum of sufficiently good quality for measurement in about 3 hours.

We now have available special *image-conversion tubes,* which intensify each image by a factor 10 or greater before it is finally recorded. Such equipment is becoming standardized and is coming into general use for radial-velocity measurements. The gain in speed more than offsets the slight loss of precision in the radial velocities obtained with an image-tube attachment. With the 200-inch reflector and an image-conversion tube, spectra with a dispersion of 128 angstrom units per millimeter are now obtainable for eighteenth-magnitude stars with exposure times of 5 hours. New digitized measuring engines help to increase the precision of measurement of the radial velocities from the spectrum photographs. The task of the person who measures the plate is simplified and the speed of recording is increased.

Photographic recording has been the traditional means of obtaining a spectrum suitable for radial-velocity measurements. Photoelectric techniques show great promise for future development. One method—pioneered by Griffin in Cambridge, England—is to use a mask that carefully images the principal spectrum lines of a star and to shift the mask laterally for minimum response, which marks the position at which the absorption lines are centrally located in their mask positions. Instantaneous measurement of the star's radial velocity at the telescope is then possible. Griffin used a 234-aperture mask to mimic the absorption lines in the spectrum of Arcturus, and he finds it possible to measure radial velocities of K stars (with spectra not unlike that of Arcturus) to the ninth magnitude with errors of ± 2 kilometers per second, using the Cambridge 37-inch reflector. There are many possible variations of the Griffin technique. A related, but different, approach is rapid scanning of small portions of the spectrum.

Thus far we have dealt only with the measurement of radial velocities for one star at a time. Beside the spectrum of each star, we impress on our photograph a comparison spectrum, produced by a luminous source mounted near the slit of the spectrograph. This gives the necessary reference lines against which we can measure the shifts of the lines in the star's spectrum. It would speed up the determination of radial velocities of faint stars if we could derive from one single photograph the radial velocities of a considerable number of stars. Numerous attempts have been made to obtain radial velocities *en masse* from objective-prism plates, which record many stars simultaneously. The hardest problem is to obtain some reference point against which to measure the shift due to radial velocity. One approach, originally developed by Pickering and improved by Bok and McCuskey at Harvard Observatory, was to use a liquid absorption filter, a vessel filled with a solution of neodymium chloride and placed in front of the photographic plate. In each spectrum a molecular absorption band is produced, which closely resembles a stellar absorption line. It serves as the desired reference zero for radial-velocity measurements. To date, the absorption method has not given results of the desired accuracy, but the basic technique is capable of further development. Good results have been obtained by Fehrenbach at the Haute Provence Observatory and at the European Southern Observatory with a scheme in which the objective prism is rotated through 180° several times in the course of one exposure. The result is that each star has two images, one with the violet end of the spectrum to the left, the other with the violet to the right. The distance between the two images of the same spectrum line in a star can be used to derive the displacement due to the star's radial velocity. The Fehrenbach technique is already providing useful material relating to the radial velocities of faint stars.

Cooperation in Research

The tradition of cooperation in research is deeply rooted in the astronomical profession. There are many reasons why this is so. First, the number of research astronomers in the world is not large (hardly more than 2,500) and there is so much work to be done that some sharing of effort is necessary to achieve common research goals. Second, we are all dealing with the same physical universe, and some consultation between workers in each area of research activity is necessary if we wish to avoid duplication of effort. Third, it is difficult to combine results by different observers unless a special effort is made to have all observations on a comparable basis: for example, in magnitude and color measurements we prefer to have all observers accept common standards and use clearly specified color filters that permit reduction from one color system to another; in proper-motion work, we prefer to have all of our motions referred to a single established fundamental system of positions and motions. Fourth, the telescopic and auxiliary equipment is often so expensive that one institution or one nation (especially some of the smaller nations) cannot afford to build what it would like to have available for its own use, and thus cooperative ownership and management become essential.

Organized cooperative research relating to the structure and motions of our Milky Way system originated in the second half of the nineteenth century. The two most famous examples of the results of such research were the *Astrographic Catalogue*, an attempt to photograph the entire sky in zones of declination, with each participating observatory assuming responsibility for the measurement of the positions of the stars on the photographs in its assigned zone, and the *Astronomische Gesellschaft Catalogue*, primarily a German undertaking, to provide accurate meridian-circle positions for large numbers of stars. The most far-reaching effort to effect international cooperation was that initiated by Kapteyn in 1904 at Groningen in Holland—the so-called Plan of Selected Areas. Kapteyn's Plan captured the imagination of the astronomical world and as a result of his appeal we now possess for many stars in the Selected Areas accurate measurements of positions, proper motions, radial velocities, magnitudes, spectral types, and color indices.

In 1918 the need was felt for a broad international organization embodying the whole of astronomy, and, in consequence, the International Astronomical Union was established. The IAU—by which three initials the organization is known throughout the astronomical world—held its first formal meeting in 1922 in Rome; it has met since then once every 3 or 4 years, with only one longer interval between meetings, that caused by World War II. The principal continuing function of the IAU is to provide through its fifty permanent commissions a medium for international contact between workers in the same field the world over. The IAU has been responsible for the initiation of many cooperative projects and it serves as the clearing house for older programs like those mentioned above. Whenever the need arises, the IAU, with the support of UNESCO, sets up conferences with a limited number of specialists in attendance. These Symposia are generally initiated by one of the Commissions of the IAU; they are attended by 100 to 200 experts in the field (including a good representation of young workers in the area) and the papers are published in a Symposium Volume. IAU Commission 33, on Galactic Structure and Dynamics, has been active in sponsoring several of these Symposia. In addition, Commission 33 has been a cosponsor for several related Symposia, one on Interstellar Matter and Gas Dynamics, for example. One of the key features of these Symposia is that they are of a truly international character.

A more recent form of cooperation is one by which astronomers of different observatories share common research facilities. In the past decade several national observatories have been created. In the United States, the Kitt Peak National Observatory, with headquarters in Tucson, Arizona, and the National Radio Astronomy Observatory at Greenbank, West Virginia, with headquarters in Charlottesville, Virginia, are the two major developments, the first for optical astronomy, the second for radio astronomy.

At Kitt Peak National Observatory, 60 percent of the telescope time is assigned to visitors from other observatories, the remaining 40 percent reserved for research by Kitt Peak staff members. The Observatory has more than half a dozen telescopes, with apertures ranging from 16 to 158 inches (Fig. 23). These telescopes are fitted with the best of auxiliary equipment: spectrographs, photometers, polarimeters, infrared-recording equipment, and so on. Kitt Peak National Observatory has a major installation for solar research. Closely related to Kitt Peak National Observatory is the Cerro Tololo Inter-American Observatory in Chile (Fig. 22), with equipment not unlike that at Kitt Peak, and with its own 158-inch reflector under construction. Chilean and other Latin American astronomers have privileged access to this facility. All of these installations have been built with funds provided by the U. S. National Science Foundation, which also pays the cost of operation.

Also supported by the National Science Foundation is the National Radio Astronomy Observatory at Greenbank, West Virginia (Figs. 26 and 27). Its principal instrumentation is a beautiful 140-foot precision steerable radio telescope and a 300-foot meridian radio telescope precise enough for research to 11 centimeters wavelength. There is also a small array of movable antennas and various special-purpose instruments. Work will soon begin on a Very Large Array for high-resolution interferometry at radio wavelengths. It will be built at a high-altitude site in New Mexico.

The astronomers of Holland, Sweden, West Germany, France, and Belgium have combined their efforts toward the establishment of a European Southern Observatory. It was built and is operated jointly by the five nations and it is also located in Chile. The largest telescope under construction for this Observatory is a 140-inch reflector. Australia and Great Britain have under way a joint effort for the construction and installation of a 150-inch reflector, to be located on Siding Spring Mountain in New South Wales, Australia; a large British Schmidt telescope is already in operation at the site.

Comparable efforts of collaboration between observatories are now in effect in many parts of the world. In the United States, the University of California astronomers at Berkeley, Santa Cruz, La Jolla, and Los Angeles jointly operate Lick Observatory on Mount Hamilton, with a 120-inch reflector as their major telescope. Universities with high-quality astronomy departments located in cli-

mates that are not really suitable for astronomical observations often put their more powerful instruments at good sites, either within or outside the borders of their country. Leiden Observatory has long had a southern station at Hartebeestpoortdam in the Transvaal, South Africa, the British have maintained observatories in South Africa, and the Uppsala Observatory has had its Schmidt telescope in Australia at Mount Stromlo Observatory. In the area of radio astronomy two major national ventures are doing much toward fostering international cooperation. The 100-meter radio telescope of the Max Planck Institute near Bonn, Germany, has many astronomers from other nations on its staff, and the Westerbork Array in Holland, a mile-long string of 12 interconnected radio telescopes, is used by radio astronomers of many nationalities. There are many cooperative ventures in the Soviet Union, the best known of which is the newly established Soviet Observatory in the Caucasus, with a 250-inch reflector as its primary instrument. Small wonder that astronomers are good world travelers and fine ambassadors of international good will!

3
The Sun's Nearest Neighbors; Stellar Populations

How does the Sun rank among the stars? Is it very brilliant, average, or somewhat dull? As students of the Milky Way, we are interested in the answers to these questions not so much because we care about the Sun itself, but rather because we wish to know what varieties exist among the stars and just what proportions there are of different kinds of stars in a sample of space, and how our sample compares with other parts of the Milky Way.

The Brightest Stars and the Nearest Stars

We shall consider first two lists of stars. The first (Table 1) includes the bright stars that we know by name, together with some bright far-southern ones that cannot be seen from northern latitudes. These stars show a wide range in color, from blue Rigel and Spica, yellow Capella, and orange Arcturus to red Betelgeuse and Antares. For all these stars,

values of the parallax and distance are available, but for the more distant ones the values are uncertain; their trigonometric parallaxes are of the same size as the unavoidable errors of measurement.

Figure 40 shows how the brightest stars vary in spectral class and in absolute magnitude. All spectral classes are present, but we note that 11 out of the 20 are the very hot B or A stars. All but one of these are more luminous than the Sun, with its visual absolute magnitude of +4.8. In fact, Rigel may shine as brightly as 52,000 Suns and five others are 1,000 or more times as bright as the Sun.

The star that comes closest to being a twin of our Sun in spectrum and luminosity is our nearest neighbor, Alpha Centauri, the brighter component of a visual binary star. It is only 1.33 parsecs away. Deneb, which is probably the most distant star in Table 1, is at about 450 parsecs, a distance too great for the accurate measurement of its parallax; the

Table 1. The twenty brightest stars.

No.	Name	Visual apparent magnitude, V	Spectral luminosity class	Parallax (arcsec)	Visual absolute magnitude, M_V	Distance (parsec)	Visual luminosity (Sun = 1)	Remarks
1	Sirius	−1.47	A1V	0.375	+1.4	2.7	23	(1)
2	Canopus	−0.73	F0Ib	.018	−3.0	56	130	
3	Alpha Centauri	0.33	G2V	.751	+4.7	1.33	1.1	(2)
4	Vega	.04	A0V	.123	+0.5	8.1	52	
5	Arcturus	.06	K2IIIp	.090	−0.2	11.1	100	
6	Rigel	.08	B8Ia	—	−7	250	52,000	(3)
7	Capella	.09	G8III + F	.072	−0.6	13.7	145	(4)
8	Procyon	.34	F5IV	.288	+2.6	3.5	7.6	(5)
9	Achernar	.47	B5IV	.023	−2.7	44	1,000	
10	Beta Centauri	.59	B1II	.016	−3.4	62	1,900	(6)
11	Altair	.77	A7V	.198	+2.3	5.1	10	
12	Betelgeuse	.80	M2Iab	—	−5	150	8,300	(7)
13	Aldebaran	.86	K5III	.048	−0.7	20.8	160	(8)
14	Spica	.96	B1V	.021	−2.4	48	760	(9)
15	Antares	1.08	M1Ib	.019	−2.5	53	830	(10)
16	Pollux	1.15	K0III	.093	+1.0	10.8	33	
17	Fomalhaut	1.16	A3V	.144	+2.0	6.9	13	
18	Beta Crucis	1.24	B0.5IV	—	−5	175	8,300	
19	Deneb	1.26	A2Ia	—	−7	450	52,000	(11)
20	Regulus	1.36	B7V	.039	−0.7	25.8	160	(12)

Remarks

(1) Sirius possesses a white-dwarf companion, V = 8.5.

(2) Alpha Centauri is a triple system. The brightest star is accompanied by a companion, V = 1.5, K6. Proxima Centauri, V = 10.7, M5eV, is the third member of the system.

(3) The trigonometric parallax of Rigel is not measurable; the listed values of M_V, distance, and luminosity are approximate. Rigel is a spectroscopic binary and has a binary star, V = 7.6 and V = 7.6, SpB5, nearby.

(4) Capella is a triple system, with companions, V = 10.0, M1, and V = 13.7, M5; Capella itself is a spectroscopic binary.

(5) Procyon itself is a spectroscopic binary. It is a triple system consisting of a close companion with V = 13.5, and a second companion with V = 12.2. The close companion is a white dwarf, spectral type F.

(6) Beta Centauri is a spectroscopic binary and has a faint component, V = 8.7.

(7) Betelgeuse is a semiregular variable and a spectroscopic binary. Its trigonometric parallax is not measur-

able; the listed values of M_V, distance, and luminosity are approximate.

(8) Aldebaran has a dM2 component, V = 13.5.

(9) Spica is variable and is a spectroscopic binary.

(10) Antares is a semiregular variable. It has a dwarf B4V companion, V = 5.

(11) The trigonometric parallax of Deneb is not measurable; the listed values of M_V, distance, and luminosity are approximate values only.

(12) Regulus has a dK1 companion, V = 7.9, and possibly two fainter companions.

Notes

1. Alpha Crucis is not listed with the twenty brightest stars, since it is a beautiful optical double, with one component, V = 1.58, B1IV, the other V = 2.09, B3n, both stars too faint for inclusion in our list.

2. The data for Table 1 are all taken from the Yale *Catalogue of Bright Stars* (third ed., 1964).

3. In the calculation of the visual luminosities, the visual absolute magnitude of the Sun has been taken to be +4.8 and visual apparent magnitude −26.8.

40. The twenty brightest stars. The diagram is
based on the data of Table 1 and shows the visual
absolute magnitudes plotted against spectral type.
(Data from the Yale *Catalogue of Bright Stars*.)

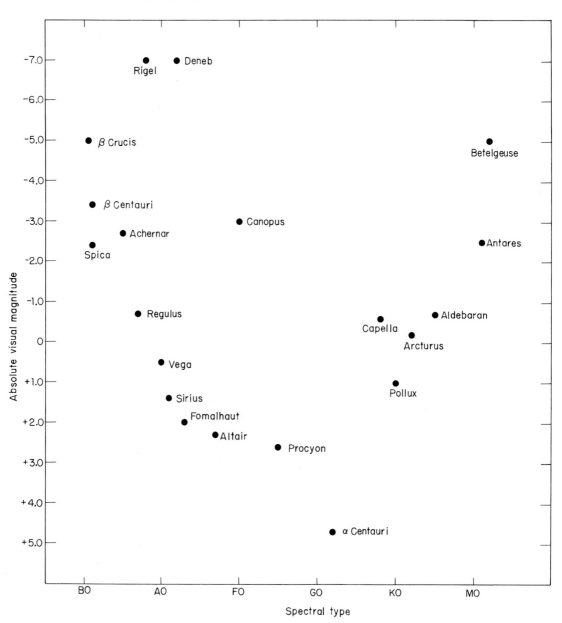

same applies to Rigel, Betelgeuse, and Beta Crucis. If we consider the volume of space that we have covered before we reach a star such as Deneb, we see that it is some $(450/1.33)^3$ or about 40,000,000 times as large as that we would have to explore to find Alpha Centauri. We may catch a minnow in our hands close to shore, but we must sail far if we wish to harpoon a whale! We begin to suspect that such distant stars as the blue supergiants Rigel and Deneb must be very rare objects in space as compared with our Sun and stars like it.

The M-type star Betelgeuse is also a very luminous star, but it is closer to the Sun than Rigel. Since its surface is comparatively cool, it must be very large to give off so much light. Betelgeuse and Antares are among the very few stars for which the diameter can be measured; the instrument used is called an interferometer. The diameter of Betelgeuse has been found to be about 600 times that of the Sun; that of Antares is only slightly less. A star so big that Mars could move in its orbit around the Sun inside that star is indeed a giant! Betelgeuse is variable in brightness and apparently also changes in size with an irregular period.

Let us next look at the list of nearby stars in Table 2. Included are all stars that are known to be within a distance from the Sun of 5 parsecs. We note that four of the brightest stars in this list—Sirius, Altair, Procyon, and Alpha Centauri—are also in Table 1. They are conspicuous stars in our sky because they are nearby rather than because of their exceptional intrinsic luminosity. The rest of the stars are much fainter, both apparently and absolutely. Figure 41 shows how these stars rate in spectral class and in absolute magnitude.

Our first list is complete, but we cannot be sure that the same is true of our second list. A few stars have been added in the past 20 years and a certain amount of reshuffling has taken place as more accurate values of the parallaxes have been determined. No doubt there will be more additions in the future. There are many faint stars with large proper motions for which parallaxes have not yet been measured, and in time we shall undoubtedly add more stars to the number of our faint neighbors. It is not believed, however, that the number will be increased very greatly. From the average speed of the stars near our Sun, the average total density of matter in our region of space can be estimated. This value, which is a little above 0.1 solar mass for a cube 1 parsec on a side, is very close to the value of the average space density that we find in the sphere with a radius of 5 parsecs, if we make reasonable assumptions as to the masses of the nearby stars and take account of the contributions from the interstellar gas and dust.

There are some very real differences between the kinds of stars on our two lists. The first list contains almost entirely what are known as giants and supergiants. They are of all spectral classes from B through M. Our second list comprises the dwarfs, or, as we prefer to call them, the "main-sequence stars." We see in Fig. 41 the very definite tendency for the stars to fall along a diagonal line, the faintest ones being the reddest; this diagonal line is called the *main sequence*. We find no stars of class K or M that are of the same absolute magnitude as our Sun. We see by an inspection of Figs. 40 and 41 that K and M stars are either intrinsically brighter than our Sun or very much fainter. We clearly see a division into giants and dwarfs. Among the nearby stars (Table 2), we find none of the brilliant B stars or G, K, or M giants. These

Table 2. Stars nearer than 5 parsecs.[a]

No.	van de Kamp no.	Name	Visual apparent magnitude	Spectral class	Parallax (arcsec)	Visual absolute magnitude	Visual luminosity	Distance (parsec)
1	1	Sun	−26.8	G2		4.8	1.0	—
2	2	Alpha Centauri A	0.1	G2	0.760	4.5	1.3	1.32
3	3	Alpha Centauri B	1.5	K6	.760	5.9	0.36	1.32
4	4	Alpha Centauri C	11.	M5e	.760	15.4	.00006	1.32
5	3	Barnard's star[b]	9.5	M5	.552	13.2	.00044[b]	1.81
6	4	Wolf 359	13.5	M8e	.431	16.7	.00002	2.32
7	5	BD + 36° 2147[b]	7.5	M2	.402	10.5	.0052[b]	2.49
8	6	Sirius A	−1.5	A1	.377	1.4	23.	2.65
9	6	Sirius B	8.3	DA	.377	11.2	0.0028	2.65
10	7	Luyten 726-8A	12.5	M6e	.365	15.3	.00006	2.74
11	7	Luyten 726-8B	13.0	M6e	.365	15.8	.00004	2.74
12	8	Ross 154	10.6	M5e	.345	13.3	.0004	2.90
13	9	Ross 248	12.2	M6e	.317	14.7	.00011	3.15
14	10	ε Eridani	3.7	K2	.305	6.1	.30	3.28
15	11	Luyten 789-6	12.2	M6	.302	14.6	.00012	3.31
16	12	Ross 128	11.1	M5	.301	13.5	.00033	3.32
17	13	61 Cygni A	5.2	K5	.292	7.5	.083[b]	3.42
18	13	61 Cygni B	6.0	K7	.292	8.3	.040	3.42
19	14	ε Indi	4.7	K3	.291	7.0	.13	3.44
20	15	Procyon A	0.3	F5	.287	2.6	7.6	3.48
21	15	Procyon B	10.8	—	.287	13.1	0.0005	3.48
22	16	Σ 2398 A	8.9	M4	.284	11.2	.0028	3.52
23	16	Σ 2398 B	9.7	M5	.284	12.0	.0013	3.52
24	17	BD + 43° 44 A	8.1	M1	.282	10.4	.0058	3.55
25		BD + 43° 44 B	11.0	M6	.282	13.13	.00040	
26	18	CD − 36° 15693	7.4	M2	.279	9.6	.012	3.58
27	19	τ Ceti	3.5	G8	.273	5.7	.44	3.66
28	20	BD + 5° 1668	9.8	M4	.266	11.9	.0014[b]	b 376
29	21	CD − 39° 14192	6.7	M1	.260	8.8	.025	3.85
30	22	Kapteyn's Star	8.8	M0	.256	10.8	.0040	3.91
31	23	Kruger 60 A	9.7	M4	.254	11.7	.0017	3.94
32		Kruger 60B	11.2	M6	.254	13.2	.00044	3.94
33	24	Ross 614 A	11.3	M5e	.249	13.3	.0004	4.02

Table 2 (continued).

No.	van de Kamp no.	Name	Visual apparent magnitude	Spectral class	Parallax (arcsec)	Visual absolute magnitude	Visual luminosity	Distance (parsec)
34		Ross 614 B	14.8	—	.249	16.8	.00002	4.02
35	25	BD − 12°4523	10.0	M5	.249	12.0	.0013	4.02
36	26	Van Maanen's Star	12.4	DG	.234	14.2	.00017	4.27
37	27	Wolf 424 A	12.6	M6e	.229	14.4	.00014	4.37
38		Wolf 424 B	12.6	M6e	.229	14.4	.00014	4.37
39	28	G158 − 27	13.8	M	.226	15.5	.00005	4.42
40	29	CD − 37° 15492	8.6	M3	.225	10.4	.00058	4.44
41	30	BD + 50°1723	6.6	K7	.217	8.3	.040	4.61
42	31	CD − 46°11540	9.4	M4	.216	11.1	.0030	4.63
43	32	CD − 49°13515	8.7	M3	.214	10.4	.0058	4.67
44	33	CD − 44°11909	11.2	M5	.213	12.8	.00063	4.69
45	34	Luyten 1159 − 16	12.3	M8	.212	13.9	.00023	4.72
46	35	BD + 15°2620	8.5	M2	.208	10.1	.0076	4.81
47	36	BD + 68°946	9.1	M3.5	.207	10.7	.0044[b]	4.83
48	37	L145 − 141	11.4	—	.206	12.6	.0008	4.85
49	38	BD − 15°6290	10.2	M5	.206	11.8	.0016	4.85
50	39	40 Eridani A	4.4	K0	.205	6.0	.33	4.88
51	39	40 Eridani B	9.5	DA	.205	11.2	.0027	4.88
52	39	40 Eridani C	11.2	M4e	.205	12.8	.00063	4.88
53	40	BD + 20°2465	9.4	M4.5	.202	10.9	.0036[b]	4.95
54	41	Altair	0.8	A7	.196	2.3	10.	5.10
55	42	70 Oph- iuchi A	4.2	K1	.195	5.7	0.44	5.13
56		70 Oph- iuchi B	6.0	K6	.195	7.5	.083	5.13
57	43	AC + 79°3888	11.0	M4	.194	12.4	.0009	5.15
58	44	BD + 43°4305	10.1	M5e	.193	11.5	.0021[b]	5.18
59	45	Stein 2051 A	11.1	M5	.192	12.5	.0008	5.21
60		Stein 2051 B	12.4	DC	.192	13.8	.0003	5.21

Note: D indicates a white dwarf; DA a white dwarf of spectral type A.

[a] Based on P. van de Kamp, *Annual Review of Astronomy and Astrophysics* 9 (1971), 104–105, Table 1.

[b] Has an invisible companion.

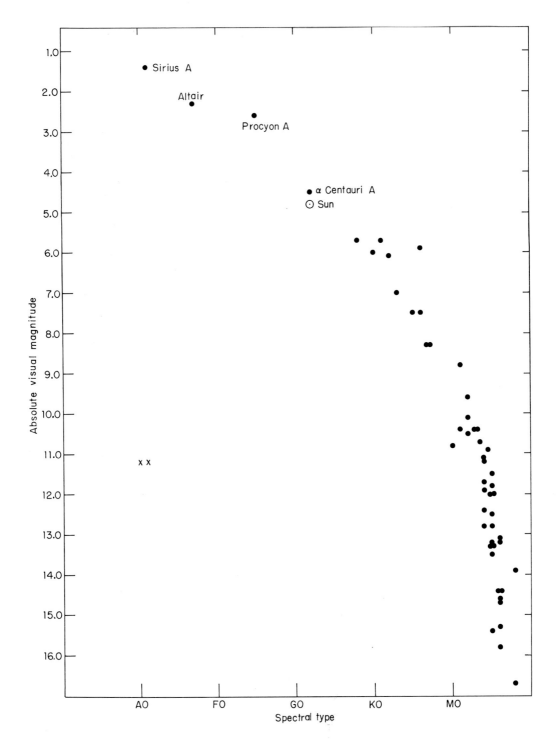

41. The nearest stars. Table 2 is the source of data for this diagram, in which visual absolute magnitude is plotted against spectral type. Only the two white dwarf A stars (DA) are plotted. (Data from van de Kamp.)

show up among the naked-eye stars because of their high intrinsic luminosities, but actually they are very rare objects in space, and those that we see are all well beyond our 5-parsec sphere.

The most common variety of star in Table 2 is the faint red dwarf of class M. These stars make up almost two-thirds of the list of our neighbors and range in luminosity from about 1/100 to 1/60,000 that of the Sun. Van de Kamp has pointed out that the faintest known star, which is less than one-tenth as bright as the faintest star in our list, lies outside the 5-parsec sphere. It is van Biesbroeck's star, BD $+40°4048$, with an estimated visual absolute magnitude $M_v = +18.6$, spectral class M5, visual luminosity 1/300,000 that of the Sun, and a distance from the Sun of about 6 parsecs. We may reasonably expect that there will be additions to our list at the tail end of the line.

Many of the M dwarfs are known as "flare stars." Their usual luminosities are very low, but they may on occasion brighten by 2 magnitudes or more for brief intervals, and several have repeated this flare-up more than once. At least five stars in Table 2 are flare stars. Bright lines are observed in their spectra, which are classified with a letter e added after their assigned spectral class.

We should mention that our list of 44 nearby stars (including the Sun) is really a list of 44 systems. Eleven of these 44 stars are double, and two are triple. In addition there are seven stars with as yet unseen companions. These invisible companions are indicated by the perturbation in proper motion of the visible star. Their masses begin to approach the range of planetary masses as we know them in our solar system and are of the order of a few hundredths of a solar mass; the largest planet, Jupiter, has a mass 0.001 that of

the Sun. The companion of Ross 614A has been found by Miss Lippincott to be a star of very low mass, one-twelfth that of the Sun. Luyten has found a double star for which the mass of each component is even smaller, probably not more than one twenty-fifth the mass of the Sun. The gap between stars and planets seems to be closing!

Our list of the nearest stars contains six blue-white stars of very low intrinsic brightness, the representatives of the class of *white dwarfs*. These stars constitute a most interesting group among our neighbors; the most famous white dwarf is the companion of Sirius. Two others are companions to bright stars, and two are single. When the companion of Sirius was discovered, its high temperature combined with its low luminosity suggested a most unusual object and probably a very rare one. No one had previously considered the possibility of the existence of stars with masses only slightly less than that of the Sun but with radii hardly larger than that of the Earth. In a recent tabulation of stars within 20 parsecs of the Sun, Gliese lists 49 white dwarfs. The searches of Luyten and others have shown that white dwarfs are as common objects as stars like our Sun. Altogether, Luyten has identified about 3,000 "certain, probable, and possible" white dwarfs, which he calls "the easiest stars to identify and the hardest to observe." The criteria for their discovery are a large proper motion and a color index comparable to that of an unreddened B or A star.

The bulk of the known white dwarfs are of about fourteenth apparent magnitude. For more than half of those discovered, spectra and parallaxes have been determined, so that they can be fitted into a spectrum–absolute-magnitude array. They form a progression not quite parallel to the main sequence, with luminosities from $M_v = +10$ to $+15$, hence 0.01 to 0.0001 the brightness of the Sun. For the stars that are strictly white dwarfs, the range in color index $B - V$ is between -0.1 and $+0.6$. Several are found as companions to other stars, so that their masses can be determined and their sizes estimated. Most of them seem to range in mass between 0.1 and 1 solar mass. Their sizes fall between the diameter of the planet Mercury and that of the planet Uranus, hence one-third to four times the size of the Earth. Luyten has recently estimated that 5 percent of all stars are probably white dwarfs.

The Hertzsprung-Russell Diagram

The discussion of the brightest and the nearest stars has given us a good idea of the kind of stars that exist in space. Figures 40 and 41 show the relation between absolute magnitude and spectral class for the brightest stars and the nearest stars. Such arrays are known as Hertzsprung-Russell diagrams, carrying the names of two of the great astronomers of our time.

Figure 42 shows a schematic drawing of the diagram with the main sequence and the giant branches drawn in. This diagram is one that applies to the stars that are thought of as being among the Sun's neighbors, taken in the larger sense, and spoken of as Population I. These stars are known to be characteristic of the stellar population in the arms of the spiral galaxies, and most of the varieties occur also in the spaces between the arms; Population II, which has a very different spectrum-luminosity diagram, is characteristic of the outer halo and of the inner parts of a galaxy, the sections where no spiral structure is present, and is especially found in elliptical galaxies. The division into Population I and Population II seems to be a very significant

distinction when it comes to showing how spectral characteristics and peculiarities of motion are related, and considerations of populations will be vital when we come to discuss the origins of stars, their probable evolution, and their ultimate fates. We note that Population I contains very bright B stars, supergiants of all classes, main-sequence stars, and the white dwarfs.

Figure 42 includes one new variety of stars, the *subdwarfs*. These are mostly Population II dwarfs of an extreme variety, whose atmospheres are almost pure hydrogen, with an unusually small percentage abundance of the heavier elements, the metals. These are the stars that presumably originated for the greater part in the earliest stages of cosmic evolution.

The diagram indicates only the mean values of the absolute magnitudes of the stars. There will be a certain spread about the mean values of these absolute magnitudes, but in general the stars conform reasonably well to the rules set up by the majority.

The Sun's Motion

There is excellent evidence to show that our Sun moves with a speed of about 18 to 20 kilometers per second with respect to the stars within a distance from it of 100 parsecs or so. The evidence for this solar motion appears clearly if we examine the radial velocities of the naked-eye stars. Figure 43 shows a projection of the sky, so drawn that equal areas on the sphere are equal areas on the paper. The sky has been divided into 94 equal areas. For each of these regions the available radial velocities of the naked-eye stars have been averaged. Altogether, the radial velocities of 2,149 stars are used, so that in each area there are on the average 20 to 25 stars.

If the Sun were at rest and if the stars were

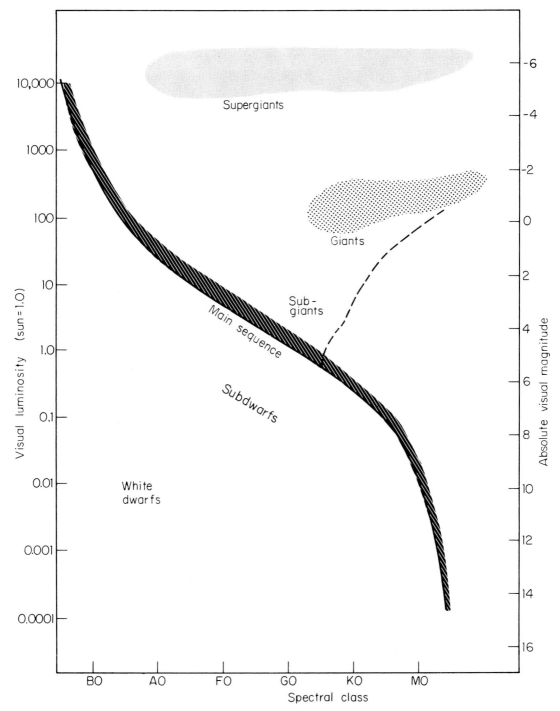

42. The Hertzsprung-Russell diagram for stars in the neighborhood of the Sun. The visual absolute magnitude (right-hand scale) or the visual luminosity (left-hand scale) is plotted against spectral class. Such a diagram was first plotted by H. N. Russell in 1913. (From Lawrence H. Aller, *Atoms, Stars, and Nebulae*, revised edition, 1971.)

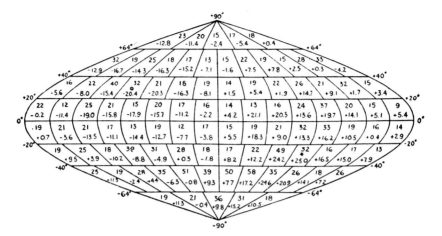

43. The Sun's motion from radial velocities. Averages of the radial velocities for 2,149 naked-eye stars measured at the Lick Observatory. The position of the apex of the Sun's motion is shown by a small circle near the position of greatest average negative radial velocity.

moving at random, there should be roughly as many positive as negative values for the radial velocities and the resultant average should be close to zero in each of the areas. Figure 43 shows that this is not so. The stars near the circle in the upper left-hand part of the figure have an average radial velocity of −20 kilometers per second; those near the asterisk in the lower right-hand part average +20 kilometers per second. Since a negative value indicates approach, it would seem that, as viewed from the Sun, all the stars in one part of the sky are marching toward us; in the opposite region they are moving away.

With reference to the naked-eye stars, the Sun is moving toward a point in the constellation Hercules, not far from the bright star Vega, at the rate of 20 kilometers per second. The circle in Fig. 43 is the *apex* of the sun's motion; the asterisk is the *antapex*.

At the rate of 20 kilometers per second, in the course of a year (about 31,600,000 seconds) the Sun will travel 630,000,000 kilometers, or the equivalent of 4.2 astronomical units. Our Earth moves steadily along with the Sun at the same rate. After an interval of 25 years we are more than 100 astronomical units from our starting place. We can take sights on the stars as we march along and measure their average displacements.

Let us see how the proper motions of the stars are affected by the Sun's motion. Figure 44 is the same projection of the sphere that we had in Fig. 43 but this time we have chosen it to assist us in examining the proper motions of 726 A stars of the fifth apparent magnitude. They are divided into 42 groups according to their positions in the sky, and for all groups we have determined the average proper motions, which are shown by the lengths of the arrows and the directions in which they point.

You will notice that most of the arrows seem to be directed away from the solar apex and toward the solar antapex. With respect to the Sun, the stars seem to be moving toward

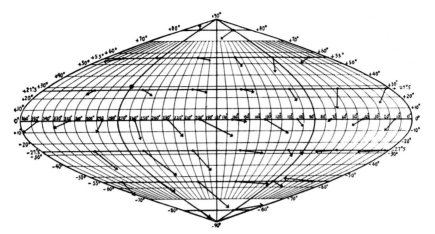

44. The Sun's motion from proper motions. The arrows represent the average directions and amount of proper motions for 726 A stars of the fifth apparent magnitude.

the antapex. What is really happening is that the Sun is moving in the opposite direction with respect to the A stars.

Since proper motions are angular displacements on the sky, they will tend to be largest along the circle at right angles to the direction in which we are traveling. In Fig. 44 they should be greatest along the projection of the circle that falls halfway between the apex and the antapex, and this is indeed observed. The average of these maximum lengths is of the order of 0.040 second of arc per year for the A stars in Fig. 44, that is, for the A stars of fifth apparent magnitude.

Mean Parallaxes

We may assume that the A stars of fifth apparent magnitude yield a solar motion from radial velocities not unlike that shown by the mass of the bright stars (Fig. 43), that is, of the order of 20 kilometers per second toward the vicinity of Vega. Hence we know that, at the average distance of the fifth-magnitude A stars, 20 kilometers per second corresponds to an average annual proper motion of 0.040 sec-

ond of arc. In making a comparison between the two values, we bear in mind one outstanding difference between radial velocities and proper motions. Radial velocities do not depend on the distances of the stars. So long as a star is bright enough to appear on a spectrum plate, its radial velocity can be determined and it matters not at all whether the star is nearby or distant. Proper motion, on the other hand, varies with the distance, growing smaller for a given linear velocity as distance increases. The effect of the solar motion on the stars will therefore depend on the average distance of the group of stars under investigation. The effect on average proper motion will be largest for the nearby stars. You see now why we chose the proper motions of a group of A stars all of one apparent magnitude. For the A stars there is no division into giants and dwarfs, such as occurs in later-type stars, so that all the A stars of the fifth magnitude will be at about the same distance. What is the average distance for the A stars of fifth apparent magnitude?

By definition, the parallax of a star is the

angular displacement corresponding to one astronomical unit at the distance of the star. We have an angular displacement of 0.040 second of arc per year, which corresponds to 20 kilometers per second or 4.2 astronomical units per year for the group of stars. The mean parallax of our A stars is therefore $0.040/4.2 = 0.0095$ second of arc and their average distance is of the order of 105 parsecs.

This distance is well beyond the distance for which accurate trigonometric parallaxes can be obtained. We shall have to remember that it is only an average distance and that it may be considerably in error for an individual A star. But it is a reliable average and we can go one step further and compute from it the corresponding average absolute magnitude for our A stars. For a star of apparent magnitude $m = 5.5$ at a distance $d = 105$ parsecs, the absolute magnitude can be computed from the formula $M = m + 5 - 5 \log d$, which gives in our special case $M = +0.4$ as the mean absolute magnitude for our A stars. The average A star in our sample is intrinsically about 60 times as luminous as our Sun.

We are really solving two major problems through an analysis of the radial velocities and proper motions of the fifth-magnitude A stars. First, we show from the radial velocities of these stars that they exhibit the effects of the standard solar motion in radial velocities, and we can check on the direction of the Sun's motion with respect to the average of our A stars from an analysis of their proper motions. Second, we can derive from the material a mean parallax, or mean distance, for the A stars of fifth apparent magnitude, and follow this up by deriving the mean absolute magnitude for the stars under investigation.

Our method of measuring *mean parallaxes* can be applied to any group of stars with known proper motions, provided that the stars are evenly distributed over the sky. The method is still applicable for groups of stars that have an average proper motion as small as 0.010 second of arc per year and that are, therefore, four times as far away as the A stars in our special example. Radial velocities are readily obtainable for enough stars of any particular group to allow us to check that the solar motion agrees with the usual value found from other groups of stars. If all is well, we can immediately combine the results and compute a mean parallax, a mean distance, and a mean absolute magnitude. Our method will not, however, produce correct results if, unfortunately, we were to select a group that has peculiar systematic motions of its own.

The method of mean parallaxes has one great advantage over the basic trigonometric method: the total displacement from which the mean parallax is found increases with time. By waiting a longer interval, we can obtain increasingly more accurate values for the proper motions and so more reliable mean parallaxes. If the stars are so distant that we do not get a measurable effect in 10 years, we can wait 50 years. With the aid of trigonometric parallaxes we cannot reach beyond distances of the order of 50 parsecs, but through the use of the mean-parallax method we can gather information that is still reasonably accurate for average distances up to 400 parsecs or more from the Sun. Research on mean parallaxes continues to be of importance for extending our basic scale of distances. The most extensive and significant early researches in this area were done by A. N. Vyssotsky and Emma Williams Vyssotsky.

Luminosity Functions

We can see from the comparison of the brightest stars and the nearest stars that from these alone we cannot satisfy our desire for completeness in both numbers and kinds of stars. In our tiny bit of space within a radius of 5 parsecs we are fairly complete as to total numbers. But we are totally lacking in the bright B stars and the red giants that are so conspicuous among the brightest stars. If we go to a sufficient distance to include at least some of these, we need such a large volume of space that we are far from having complete information as to just how many other stars it contains.

Yet the astronomer wants to know how many stars of given absolute magnitude and spectral class are on the average present in a given volume of space. It lies within our power to count the stars to given limits of apparent magnitude. But counts alone are not enough. The astronomer does not want to see the sky as primitive people see it, as a curved surface on which bright lights appear. What we must add is the third dimension, so that we can see how the stars are spread out in space. We obviously wish to obtain the complete tabulation of the absolute magnitudes for a typical sample volume in the Milky Way, at least for the parts of the system in which our Sun is located. This tabulation, which lists the numbers of stars per unit volume for successive intervals of absolute magnitude, is known as the *luminosity function*.

It is clear that the problem of deriving the luminosity function will have to be tackled piecemeal. We can build up the general luminosity function by studying first the separate tabulations for different spectral classes and then putting them together in true proportions to obtain the total picture. We shall naturally be very curious to find out whether the same mixture of spectral classes holds in different parts of the Milky Way. But from stars with classified spectra, we shall principally obtain data on the stars intrinsically more luminous than our Sun and we know already that the great majority of stars are in-

trinsically fainter than the Sun and that mass spectral classification alone will not do the trick. We shall consider first the faint end of the luminosity function, after that the bright end, and then combine the two.

Proper motion is a powerful tool for the selection of the nearer stars, those with measurable trigonometric parallaxes, from among the stars at large. A star at a distance of 4 parsecs and moving at a rate of 20 kilometers per second of linear crosswise motion will have an observed annual proper motion close to 1 second of arc; about the same proper motion would be observed for a star at 6 parsecs distance moving at the rate of 30 kilometers per second. Since the linear crosswise speeds of stars range generally from 5 to 50 kilometers per second, with the majority between 20 and 30 kilometers per second, we may expect the majority of the stars with proper motions of the order of 1 second of arc to have measurable trigonometric parallaxes. Parallax observers naturally concentrate their efforts on varieties of stars that show promise of having measurable trigonometric parallaxes, so it is not surprising that parallax determinations are available for a sizable sample of all known stars with large proper motions. The sample is sufficiently large to permit the use of statistical techniques to correct for the absent stars and make the census figures complete and representative ones. The faint end of the general luminosity function is by now fairly well known from an analysis of available data on trigonometric parallaxes for stars with known proper motions in excess of 0.2 seconds of arc per year. The most complete study of the faint end of the luminosity function has been made in recent years by Luyten of the University of Minnesota.

Trigonometric parallaxes fail us miserably when we attempt to derive the luminosity function for stars with absolute magnitudes like that of our Sun and brighter. For these stars we turn to studies involving proper motions and radial velocities and also to work on the precise determination of spectral types and luminosities. Fortunately, we do not have to depend exclusively on the evidence from mean parallaxes for the stars beyond the reach of the trigonometric method. Spectrum-luminosity classification is now possible for all stars for which objective-prism spectra of not too low dispersion are obtainable. By careful inspection of the spectra, one may obtain an estimate of any star's absolute magnitude with an uncertainty that need not exceed one-half magnitude. In other words, we can take a fairly accurate star-to-star census and combine the results to derive the distribution function of absolute magnitudes—the luminosity function—per unit volume for all stars of all spectral types taken together. The resulting general luminosity function is shown in Fig. 45.

By a combination of the results of the various methods we can thus obtain reliable data on the distribution of the absolute magnitudes of the stars in the Sun's neighborhood. The pioneer investigations of Kapteyn were followed by extensive studies by Seares and van Rhijn. The method by which we determine at least the bright end of the general luminosity function, by combining the luminosity functions of successive relatively small intervals of spectral class, was first used in 1932 by van Rhijn and Schwassmann.

The bright end of the general luminosity function has been determined with precision through studies done at the Warner and Swasey Observatory. The work of McCuskey and his associates not only gives a very valuable check on the luminosity function for the region directly around the Sun, but provides useful information regarding the variations in the shape of the function for regions within 500 parsecs of the Sun in the galactic plane. McCuskey finds for absolute magnitudes −1 to +1 somewhat greater numbers than shown by the curve of Fig. 45, but on the whole there is little that can be done to improve the original results of van Rhijn and Schwassmann. In the galactic plane, there are fluctuations of the order of 30 percent in the relative frequencies of the various absolute magnitudes. The curve shown in Fig. 45 represents in a way the combination of two luminosity functions, one derived from proper motions, the other from a combination of separate functions for each spectral class. The curve shown represents the number of stars in a cube 1,000 light-years (about 300 parsecs) on a side.

Marked systematic variations are found if we consider the frequency distributions of absolute magnitudes at some distance from the galactic plane. The absolutely brightest stars—B and A, for instance—thin out far more rapidly on a proportional basis than do the G and K dwarfs, and the shape of the luminosity function at a few hundred parsecs above or below the galactic plane differs markedly from that shown in Fig. 45, the principal difference being a decided deficiency for the brighter absolute magnitudes at a height of a few hundred parsecs above or below the plane as compared with the relative numbers in the plane.

To supplement the schematic Fig. 42, we list in Table 3 the mean values of the visual absolute magnitudes and the spreads around these means (dispersions) for the stars of each spectral group. The values listed are applicable to a sample volume of space in the part of the Galaxy where our Sun is located. The values given for the main-sequence (Luminosity Class V) stars of spectral groups

45. The luminosity function. The numbers on the vertical scale are the numbers of stars in a cube 1,000 light-years on each side for the photographic magnitudes shown on the horizontal scale. (From data compiled by van Rhijn, McCuskey, Kuiper, and Luyten.)

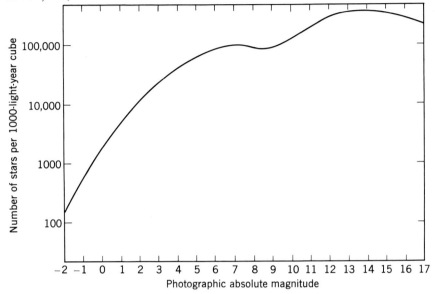

G5 and earlier and those of the subgiants (Luminosity Class IV) and of the regular giants (Luminosity Class III) are from McCuskey. We have added the standard main-sequence values for the K5V and M5V stars.

The bright end of the luminosity curve is the most important part for the study of the distribution of the stars in space. However, the dwarfs constitute the bulk of the population. Their luminosities are equal to or less than that of the Sun, so that it takes many fainter ones to shine as brightly as the giants. But their masses are not too different from the Sun's mass and these faintly shining bodies play an important role in determining the gravitational properties and the motions of the system.

Table 3. Mean visual absolute magnitudes and dispersions for different spectral groups.

Spectral group	Mean visual absolute magnitude	Dispersion around mean
B 5	−1.0	±0.5
B8–A0	+0.2	.5
A1–A5	+1.6	.4
A7–F5	+2.8	.5
F8V–G2V	+4.5	.4
G5V	+5.0	.3
K5V	+7.2	.3
M5V	+12.3	.3
F8III–K3III	+0.9	.8
F8IV–K3IV	+3.2	.8
K5III–M5III	0.0	.6

Populations I and II

If we refer to the familiar distribution of absolute magnitudes and varieties of objects near the Sun as Population I, we find a very different distribution at great distances from the Sun, especially for the Sagittarius section. Following Baade, we refer to the latter as Population II. Population I, with its abundance of intrinsically very luminous objects, is apparently characteristic of the outer parts of our own and other galaxies, whereas Population II prevails in the central cores of galaxies like our own and in the thin outer halo of our Galaxy.

Historically it is interesting to note that the realization of a separation between the two populations came not as a result of studies of our own Galaxy, but rather from Baade's researches on our nearest large neighbor galaxy, Messier 31, the famous Andromeda spiral. Messier 31 is so far from us—its distance is of the order of 650,000 parsecs—that we cannot hope to observe in it stars with absolute magnitudes comparable to that of our Sun. Even with a powerful telescope like Palomar's 200-inch Hale reflector it is impossible to detect stars as faint as $M = -1$, stars that are intrinsically 200 times as bright as the Sun. Baade reasoned that it should be possible, however, to find with relative ease anywhere in Messier 31 the stars with absolute magnitudes $M = -3$ and certainly stars with $M = -5$ and brighter. Such stars are clearly shown in the outer parts of Messier 31, where spiral structure prevails, but Baade found it impossible to photograph in normal blue light any individual stars in the central regions of Messier 31. However, he succeeded with the help of red-sensitive photographic plates in resolving the nucleus of Messier 31 and the elliptical galaxy NGC 205 that accompanies

it. The brightest stars proved to be red stars with visual absolute magnitudes between -3.5 and -4.0 and with color indices of the order of 1 magnitude and greater.

Baade concluded from these observations that the stars that constitute the nucleus of Messier 31 are very different (Population II) from those that are found in the outer spiral regions (Population I). The contrast between the two populations is not a minor one; the brightest stars of Population I are blue and have absolute magnitudes as high as -7 to -9, whereas the brightest stars of Population II are red and have absolute magnitudes of -3 at the most. Moreover, interstellar gas and dust abound in Population I, whereas in Population II gas and dust are either absent or obviously very minor constituents.

Baade's distinction between the two populations represents a major advance in the understanding of our own and other galaxies. We shall have to refer to it frequently, for not only does it possess tremendous significance for the study of the structure of galaxies, but the population approach has far-reaching consequences for problems of stellar birth and evolution.

In the present book we shall, for reasons of simplicity, refer mostly to the two basic populations. The reader should be warned, however, that the picture of two essentially different varieties of stars is an oversimplification. It is true that it seems to be supported by the spiral-galaxy data, but we are dealing there always with stars that are at least 5 absolute magnitudes brighter than our Sun; even our best equipment does not permit us to observe the common stars of moderate magnitude. The authors—and along with them quite a few of their colleagues—prefer to think in terms of at least three basic varieties of stars: Population I, responsible for the principal spectacular spiral features; the *Common Stars*, first so named by Oort, which inhabit the more or less amorphous regions between the spiral arms; and Population II, the nuclear population, which spills over to some extent into the outer parts.

Perhaps it is most fitting to think in terms of six major groupings, somewhat in the following manner:

Extreme Population I consists of the objects that show a very marked preference for association with structural spiral features. We include in our listing the interstellar gas and the cosmic dust, which are highly concentrated in the spiral arms and which provide the "building blocks" from which the new stars are formed. The stars associated with Extreme Population I are cosmically very young; the ages since formation from the interstellar gas and dust for the stars of Population I are 20 to 50 million years at the most, less than one-fifth of the period of one galactic revolution, hence less than one-fifth of a *cosmic year*. Some Population I stars have ages of less than 1 million years and some may have been caught in the act of formation, or directly after they had become stars. Population I is found only within a few hundred parsecs of the central galactic plane. It is confined to a wafer-thin galactic layer, a ring with an inner radius of 5,000 parsecs and an outer radius of 15,000 parsecs, about 500 parsecs thick.

Stars of Extreme Population I occur generally in regions where clouds of interstellar gas and dust are plentiful. The presence of *emission nebulae* (Chapter 8) is characteristic of regions with Population I; these nebulae can be found and studied by optical and by radio techniques. *Neutral atomic hydrogen* (H I) is recognized through its 21-centimeter radiation. *Cosmic grains* (solid particles) are found in dark nebulae (Chapter 9)—often with associated interstellar molecules present—and they can be detected by the reddening effects produced in the light of distant stars.

The varieties of stars associated with Extreme Population I are:

1. O to B2 stars, single, in clusters, or in associations;
2. Late-type supergiants, including especially those cepheid variables with periods in excess of 10 to 13 days;
3. Wolf-Rayet stars;
4. B emission stars;
5. T Tauri variable stars;
6. Certain infrared objects.

Normal Population I stars are principally the somewhat older stars, with ages since formation possibly as great as 2 or 3 cosmic years. These stars are old enough to have diffused away from their places of origin in spiral arms and from the structures to which they belonged originally. They are generally found close to the central plane of the Galaxy. The principal varieties of stars associated with Normal Population I are:

1. The B3 to B8 stars and the normal A stars. Open clusters with their brightest stars of these spectral types, but without O to B2 stars, are Normal Population I. These clusters do not fit well into observed spiral patterns and they have generally no associated interstellar gas;
2. Stars with strong metallic lines, spectral classes A to F;
3. The less brilliant supergiant red stars.

Disk Population stars, Oort's Common Stars, are next in line. They have cosmic ages in the range between 1 and 5 billion years, that is, between 5 and 25 cosmic years. Our Sun is one of the Common Stars. This population includes the great masses of the inconspicuous stars found mostly within about 1,000 parsecs of the central plane in a galactic belt with an inner radius of 5,000 parsecs and an outer radius of 15,000 parsecs, centered upon the galactic center. The following varieties of stars belong to the Disk Population:

1. G to K normal giant stars (Class III);
2. G to K main-sequence stars (Class V);
3. Long-period variable stars with periods greater than 250 days;
4. Semiregular variable stars;
5. Most disk planetary nebulae;
6. Novae (somewhat questionable);
7. Older open clusters.

Intermediate Population II stars are those that have considerable spreads in velocities, sufficiently high to be found farther above or below the galactic central plane than 1,000 parsecs. Whereas Population I stars and Common Stars move in nearly circular orbits around the galactic center, the Intermediate Population II stars move in quite elongated orbits around the same center. Many of these stars may have originated at distances of no more than 5,000 parsecs from the galactic center. Most of the really old stars (with ages of 50 to 80 cosmic years) may belong to this class. The principal varieties classed under Intermediate Population II are:

1. High-velocity stars;
2. Weak-line stars;
3. Long-period variable stars with periods less than 250 days, but greater than 50 days;
4. W Virginis cepheid variables;
5. RR Lyrae variable stars;
6. White dwarfs;
7. Oldest known open clusters.

Halo Population stars are those formed in the earliest stages of development of the Gal-

axy, which was then presumably much less flattened than at present. In all likelihood the building of elements heavier than hydrogen and helium had not progressed very far. The principal stars generally associated with the Halo Population are:

1. Subdwarf stars;
2. Halo globular clusters;
3. RR Lyrae stars;
4. Weak-line stars (extreme);
5. Highest-velocity stars.

Nuclear Population stars are a variety about which we know the least. The great strength of the sodium *D* lines in the spectra of the nuclei of many spiral galaxies, first noted by Spinrad, and the great strength of the cyanogen (CN) molecular bands (Morgan and Mayall) may suggest that the nuclear regions of our own and other spiral galaxies contain vast numbers of M-type dwarf stars. The stars listed below are generally considered as Nuclear Population stars, but we recognize from the start that such assignments are quite uncertain because of obvious observational limitations:

1. RR Lyrae stars;
2. Relatively metal-rich globular star clusters;
3. Planetary nebulae;
4. Vast numbers of M-type dwarf stars;
5. Giant stars (G to M) with strong cyanogen bands;
6. Infrared objects.

We note that RR Lyrae variables are listed as possibly belonging in some way to Intermediate Population II and also to the Halo and Nuclear Populations. Obviously some further sorting is in the offing!

The gradual progression from Baade's two populations to the sixfold classification we suggest is very much in line with the views advocated about 1950 by Soviet astronomers, notably Parenago and Kukarkin. They argued from the start in favor of a continuous sequence of stars grouped by their physical characteristics and motions into separate sub- systems of the Galaxy. The Soviet astronomers have all along been the strongest supporters of a minimum of three major subdivisions: a highly flattened, an intermediate, and a roughly spherical component.

4
Moving Clusters and Open Clusters

On a clear night we notice, apart from the general hit-or-miss arrangement of the stars, a few places where the stars are closely clustered and seem to belong together. The Pleiades or the Seven Sisters, the Praesepe or Beehive Cluster, the Hyades in Taurus, the double cluster *h* and Chi Persei—all of these have been known since antiquity. To these naked-eye clusters, telescopic surveys have added many more. There are two varieties of star clusters: the open clusters and the globular clusters. In the present chapter, we shall be concerned with the open clusters only.

Moving Clusters

The stars of an open cluster are close together in space; they are not merely a transitory chance arrangement. If a cluster is real and has a lasting quality, then all its stars must share a common motion. They should, therefore, move through space in parallel paths and with identical speeds. If the group covers a large area of the sky—as do, for example, the Hyades—and is near enough to possess measurable proper motions, then the arrows that represent the directions of proper motion for the individual stars in the cluster will all seem to converge to one point on the celestial sphere. We generally refer to the open clusters that are close enough to us to show such measurable proper motions as *moving clusters*. The Hyades are the prototype of a moving cluster.

Lewis Boss first detected the convergent motion of the Hyades when he was preparing his *General Catalogue* (1914) of proper motions. For the stars in this cluster the proper motions are large, and it is possible to sort out the stars that belong to the cluster from the field stars (Fig. 46). Since the distances can be measured, it is possible to build up a picture of the cluster, discover the kinds of stars it contains, and determine how closely they are packed. In 1952 Van Bueren published a

46. The convergence of the proper motions of the
Hyades cluster is shown by this diagram. All stars
brighter than the ninth magnitude belonging to
the cluster have been mapped. The sizes of the
dots are a measure of the magnitudes of the stars.
The arrows show the proper motion displacements that may be expected in the course of the
next 18,000 years. It is apparent that the stars
share a common motion in space. (This figure is
by Van Bueren, who made the study at the Leiden
Observatory.)

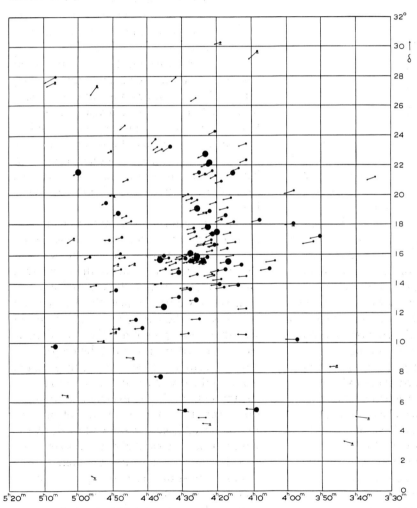

comprehensive study of the Hyades cluster, which he finds to be a flattened system with its shortest axis perpendicular to the galactic plane and about two-thirds as long as the axis in the plane. The cluster is nearby—only about 40 parsecs from the Sun—and Van Bueren lists about 350 stars as probable cluster members, from the brightest to stars intrinsically much fainter than our Sun. The Hyades members are concentrated in a somewhat irregular fashion toward the center of the cluster. The majority of the cluster members are G and K main-sequence stars. The bluest stars in the cluster are of spectral type A2 and there are a few G and K giants. In the core of the cluster the average density is at least three times that of the stars in the region around the Sun.

The Hyades moving cluster has been the subject of many investigations during recent years. Careful studies have been made of the colors and brightnesses of the recognized members, and many additions to Van Bueren's list have been published. The most recent comprehensive survey has been made by Van Altena, who has combined his new discoveries with those of Giclas and associates of Lowell Observatory and those of Luyten at the University of Minnesota. Van Altena estimates that his search is 93 percent complete for the central region of the Hyades cluster. The main sequence in the Hertzsprung-Russell diagram is very clearly marked, and there is a parallel subdwarf sequence. We now have a good list of Hyades members between visual apparent magnitudes 8 and 18, stars that have absolute visual magnitudes in the range $+5$ to $+15$. Eggen and Greenstein have found 15 white dwarf stars that are probable members of the Hyades moving cluster.

Another moving cluster that is playing an increasingly important role in galactic research is the Scorpius-Centaurus cluster. This is a moving group that is especially rich in B stars. Many of the stars that outline the magnificent arc of the Scorpion in the heavens belong to this moving group and so do most of the blue-white stars in Ophiuchus. The southern half of this great moving stream cannot be observed from northern latitudes, but it shines in all its beauty for southern observers. They see it as an isolated grouping of blue-white stars in Centaurus near galactic latitude $+10°$. Kapteyn first called attention to this moving stream in 1918, and the stream has since been the subject of many investigations. The checking on membership has progressed slowly. In recent years, Garrison has investigated the northern section and Glaspey has completed the analysis in the southern section. It appears that a good number of A stars, and some F stars, share the motions of the easily recognizable B-star members.

Moving clusters have a very special place in astronomical affairs, for it is from them that we can obtain precise basic information about absolute magnitudes of the recognized members. The principle by which this is done is really quite simple. When we plot on the celestial sphere the proper motions of the stars that we suspect of membership, then, as we have noted already, all of the proper motion vectors will seem to pass through one point on the sky, the *convergent* of the stream motion. This convergent point marks the direction toward which the stream is moving as viewed from the Sun and Earth. We have here a close parallel to the case of a meteor shower, in which all of the meteors seem to move across the heavens in paths intersecting at one single *radiant* point.

For each certain member of a moving cluster like the Hyades, we can determine the star's radial velocity (measured in kilometers per second) in addition to its proper motion. Knowing the distance in degrees between each star member and the convergent, we can deduce from these observed radial velocities the total stream motion in kilometers per second. Now, the proper motions toward the convergent are known in seconds of arc per year. For each star member, the linear velocity, in kilometers per second, corresponding to the proper motion can be readily predicted once we know the total stream motion and the position of the convergent. Hence we have for each star member both the measured proper motion in seconds of arc per year and the predicted linear velocity corresponding to it in kilometers per second. The distance of the star can thus be found. This result is most important, for it makes it possible to assign *individual* distances to all stars that are members of a moving cluster with a well-defined stream convergent. The Hyades cluster and the Scorpius-Centaurus cluster both fall in this category.

Moving clusters are of great importance to the astronomer in that they permit him to determine precise distances and absolute magnitudes for the member stars. We found in Chapter 3 that trigonometric parallaxes are excellent for purposes of calibration to distances of 20 parsecs from the Sun and Earth, but that they become increasingly less reliable for greater distances. The Hyades moving cluster, at an average distance of about 40 parsecs from the Sun, is excellently placed to yield distances for a number of stars beyond the reach of the trigonometric-parallax techniques, and the Scorpius-Centaurus cluster extends our range to well over 200 parsecs. It is especially important that, with the aid of the Scorpius-Centaurus moving cluster, we

can obtain precise distances and absolute magnitudes for individual B and A stars. These are the varieties that are very scarce within 20 parsecs of the Sun and Earth.

The System of Open Clusters

The total number of moving clusters is small, since almost all open clusters are so far away that they do not show measurable proper motions. But in spite of the absence of observable motion we can learn much about the more distant open clusters. Some are rich in numbers of stars, others are little more than slight condensations in the sky. There are some 400 known clusters that are strictly open clusters; the Pleiades cluster (Fig. 47) is probably the best known of these. The majority of open clusters are located close to or in the band of the Milky Way. There are probably many more than 400 in our Milky Way system, but the more distant clusters are not noticed against the rich stellar background along the Milky Way. Furthermore, the great majority are hidden from our view by the intervening cosmic dust close to the galactic plane. The further cataloguing of open clusters continues to be a major task facing astronomers.

Much has been learned about the distances and physical characteristics of open clusters from the classification of the spectral types of the member stars. The most significant early research in this field was done in the late 1920's by Trumpler and by Shapley. In 1930 Trumpler summarized his researches in a paper published in a *Lick Observatory Bulletin* which today stands out as a classic of its time. Both he and Shapley stressed differences in spectral composition for some of the better-known clusters, and they intimated— as since has proved to be significant—that these differences might well have importance

47. The Pleiades cluster. The Pleiades are embedded in nebulosity. (From a photograph by Barnard.)

for problems of stellar evolution. From Trumpler's study there came the first conclusive proof of the presence of interstellar absorption at low galactic latitudes. We should note here that the first relevant observations for the law of space reddening followed also from studies by Trumpler on the intensity distribution of the continuum in the spectra of some heavily obscured distant stars.

During the 1930's, it became evident that much of value was to be learned from the Hertzsprung-Russell diagrams (generally referred to in the professional vernacular as H-R diagrams) of open clusters. There are several ways in which an H-R diagram can be plotted. The first method is to plot spectral class (horizontally) against apparent or absolute magnitude (vertically). In a second method, the measured color index of each individual cluster star replaces the spectral class. Now that it is possible by photoelectric techniques to measure color indices of faint stars quickly and precisely, the second method has become the more useful of the two.

Figure 48 shows a composite H-R diagram in which the principal branches of the color-magnitude arrays are plotted for some of the best-known clusters. In the diagram we find vertically the absolute visual magnitude M_v and horizontally the color index $B - V$, the difference between the blue magnitude B and the visual magnitude V. In this diagram our Sun would be located at the point near $B - V = +0.6$, $M_v = +5$. Upon inspection, we note that the Hyades cluster falls somewhere near the middle of the diagram and that it and Praesepe are rather similar in that they have no really blue stars and no stars as bright as absolute magnitude $M_v = 0$, that is, no stars as much as 100 times as bright as our Sun. In other words, these clusters contain a few mild giant stars, but they have no

blue giants and certainly no supergiants among their members. Let us now examine some of the clusters that occupy the upper left-hand section of Fig. 48. Here we find some familiar star clusters, the Pleiades and the double cluster *h* and Chi Persei. The Pleiades cluster has a steep blue branch, with some stars reaching $M_v = -2.5$, that is, with 1,000 times the brightness of the Sun. The cluster *h* and Chi Persei outdoes all others, with blue supergiants and a few red ones at $M_v = -6$, fully 25,000 times as bright intrinsically as the Sun! But, turning to the lower half of the diagram, we find some inconspicuous clusters, like Messier 67, in which the bluest star is hardly bluer or brighter than the Sun.

There are two reasons why diagrams like Fig. 48 are of great interest to the astronomer. The first is that the information contained in them is helpful in the study of the distances of remote open clusters; the second is that from diagrams of this nature we may learn much about the ways in which the stars gradually evolve. In the present chapter, we shall concern ourselves principally with the first of these problems, saving the evolutionary implications for Chapter 11.

With the exception of not more than half a dozen nearby clusters, the distances of the remaining several hundred open clusters are too great to be measured by either the trigonometric or the moving-cluster technique. But it is possible to measure, by a combination of photographic and photoelectric techniques, the colors and magnitudes of very faint stars in open clusters. Color-magnitude arrays have now been measured for more than 100 open clusters. In every case one ends up with a diagram in which observed color (such as $B - V$ in Fig. 48) is plotted against the apparent visual magnitude V or blue

48. The Hertzsprung-Russell diagrams for the
seven galactic open clusters h and Chi Persei,
NGC 2362, the Pleiades, the Hyades, Praesepe,
NGC 752, and Messier 67 have been combined in
this diagram by Johnson and Sandage. The main
sequences of the clusters converge toward the
faint end but there are great differences in the
upper regions of the main sequence and the giant
and supergiant branches.

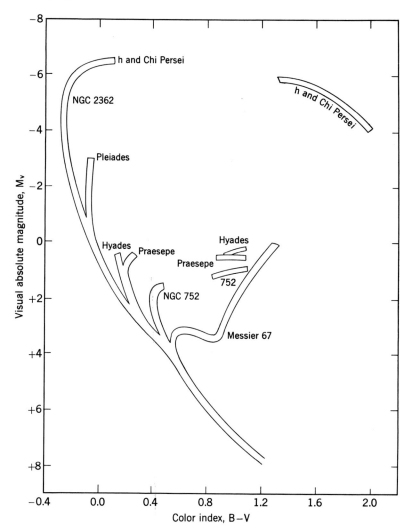

magnitude B of each cluster star. One may draw through the observed points the best-fitting curve and attempt to judge by inspection to which curve in Fig. 48 it corresponds. It is generally not difficult to decide whether the cluster one is dealing with is like the Pleiades, like h and Chi Persei, like the Hyades or Praesepe, or if perhaps it represents an extreme case, like Messier 67. In the absence of any appreciable interstellar absorption, the procedure of finding the distance to the cluster is simple and direct: one notes at various values of $B - V$ the corresponding values of M_v from Fig. 48 and the observed values of V. To check the resulting distance, one should attempt to obtain, if possible, the spectral types on the Morgan-Keenan system at least for the brightest members. From these spectral types and luminosity classes we can then readily check on the absolute magnitudes M_v of some of the individual bright stars and thus on the distance. A few well-determined spectral types and luminosity classes help immeasurably in disentangling the complex situations that often arise in practice.

Interstellar absorption and the accompanying reddening, which affects the observed colors $B - V$, complicate matters considerably, but with some care the distance problem is still amenable to relatively straightforward handling. It is simplest when the spectral types and luminosity classes are available for a few of the brighter members of the cluster. From the spectral data one can then predict for each star the intrinsic value of $B - V$ and derive the amount of reddening by noting the difference between it and the observed value of $B - V$ for the same star; we call this difference the *color excess* $E(B - V)$. From the scattering and absorption properties of the interstellar medium (see Chapter 9), we know the value of the factor by which we must

multiply the color excess to obtain the corresponding total visual absorption; for $E(B - V)$ color excesses the conversion factor is close to 3. Multiplication by this factor then yields the total photovisual absorption between the star in the cluster and the Sun and Earth. If we call this absorption A_v, then $A_v = 3 E(B - V)$ and the relation between the relevant quantities reads simply

$$M_v = V + 5 - 5 \log d - A_v.$$

In this formula we know every quantity with the exception of d, the distance of the cluster.

The great majority of the open clusters studied to date lie within 3,000 parsecs of the Sun and the remaining ones with known distances are almost all within 5,000 parsecs of the Sun. The clusters are mostly at distances less than 200 parsecs above or below the central galactic plane. In our disk-like Milky Way system open clusters are found most often in a thin layer near the central plane, the band in which—as we shall see in subsequent chapters—spiral structure prevails. Open clusters have a distribution that is totally different from that of the globular clusters (Chapter 5), which either inhabit the central regions of our Galaxy or are found to great distances from the galactic plane; globular clusters seem to avoid the outer thin layer of spiral structure. Open clusters belong for the greater part to Population I, whereas the globular clusters are of Population II.

It is obviously important that we should learn more about the distribution in space of the known open clusters with poorly determined distances. The most direct solution of this problem is through studies of colors and magnitudes. Here the photoelectric technique is the basic one but it may—and often must—be supplemented by photographic observations. The photoelectric observations need to provide only a limited number of precise standards of magnitudes and colors. We generally select the thin outer parts of a cluster for our photoelectric work. With the aid of what we may call "photographic interpolation," we can then determine colors and magnitudes for a hundred or more stars by measuring their magnitudes from photographs in two or three colors with reference to the standard stars.

There are several methods in vogue for research of this sort, one of the most effective being that developed by W. Becker of Basel, who measures his magnitudes in three carefully selected wavelength intervals and then succeeds in a straightforward manner in separating effects of space reddening from those produced by the intrinsic colors of the stars. Becker's method has yielded a detailed picture of the system of open clusters.

Wide-band photometry on the *UBV* system has been applied on a large scale to cluster studies, and intermediate and narrow-band techniques are being used increasingly in studies of moving clusters and of open clusters. It appears more and more as though the photoelectric techniques are capable of higher precision than the techniques of spectrum-luminosity classification. Not all clusters are of the same age and the chemical compositions of the atmospheres of the member stars may differ appreciably from one cluster to the next. We shall deal with the evolutionary effects in later chapters, but we should note here that astronomers, notably Harold Johnson, have obtained one basic reference line, the Zero-Age Main Sequence, which is very close to the main sequence shown in Fig. 42. It provides the basic reference line for stars of average chemical composition which, following their birth from the interstellar medium, have contracted onto the main sequence.

Once the stars, through a process of contraction from the interstellar medium, have arrived on the main sequence, they derive their energy of radiation chiefly from the continual building up of helium nuclei from hydrogen nuclei. The principal evolutionary effects that follow as these stars use up their internal hydrogen supplies of nuclear fuel is a trend toward the upper right-hand part of Fig. 48. The intrinsically brightest stars in the upper left-hand part of the figure will exhaust their hydrogen supplies most quickly and, as calculations show, they will move rapidly to the right in the diagram. Gradually the disease of hydrogen exhaustion spreads downward along the main sequence. Obviously, the clusters that still have very luminous blue-white stars (in the upper left-hand corner of Fig. 48) are the youngest, and the older clusters will show a chopped-off main sequence, with a few evolved red giants in the upper right-hand corner.

Associations and Aggregates

In addition to the tightly knit clusters of stars, there are more loosely connected distant groupings known as *stellar associations* and *aggregates.* For many years astronomers had been aware of the existence of some very extended moving clusters—the Scorpius-Centaurus cluster of B stars, for instance—recognized by their common space motions. The Russian astronomer Ambartsumian was the first to demonstrate that there exist a considerable number of loose groupings, called *associations*, most of which are too distant to show detectable parallelism of proper motions. These occur especially among the very luminous—and presumably very young—O and early B stars. The mutual gravitational attraction between the stars is too weak to hold the association permanently together, but it may have not existed long enough to have been torn asunder by the gravitational forces of the Milky Way, or to have drifted apart owing to the individual motions of the stars. These associations have been studied primarily by Ambartsumian in the Soviet Union and by Morgan at the Yerkes Observatory. The familiar Orion region, with its bright and dark nebulae and an abundance of O and B stars, has several associations, which has led Morgan to refer to it as an *aggregate*.

Every amateur astronomer is familiar with the beautiful Orion Nebula and the associated Trapezium cluster. The nebula itself is a large cloud of ionized hydrogen gas made luminous by the ultraviolet radiation emitted by the hot stars of the Trapezium cluster. In the same general region of the sky, one finds a very loose grouping of O and B stars, slightly elongated in shape, and extending over a far greater volume of space. Radio observations by Menon at the G. R. Agassiz Station of the Harvard Observatory and by others reveal that all these features are imbedded in a very large cloud of neutral hydrogen with a diameter of the order of 100 parsecs and with a total mass of the order of 50,000 to 100,000 solar masses; the famous Orion Nebula, with a total mass probably no greater than 1,000 solar masses, is just a little sore spot of ionized hydrogen in the larger complex. Many strong infrared sources are found associated with the Orion complex—probably indicative of protostar formation. Within this complex are several well-known gaseous and dust features, notably the Horsehead Nebula, a beautiful sight when photographed with the 200-inch Hale reflector (Fig. 49). One of the most striking features is a faint arc of nebulosity, first photographed by Barnard. It represents radiation from ionized hydrogen and is apparently formed at the edge of the expanding large

neutral gas cloud, possibly caused by a shockwave phenomenon that occurs as the huge gas mass (expanding, according to Menon, at a rate of about 10 kilometers per second) bumps into the surrounding interstellar matter. In an extensive study of the Orion aggregate, Parenago has found evidence of a rotation of the entire system of stars and nebulosity. The radio observations of Menon confirm the presence of this rotation. We thus see the Orion aggregate as an enormous interstellar boiling pot, mostly of neutral hydrogen gas. Some of it apparently is condensed into relatively young stars, which ionize some of the gas and cause it to shine. Interstellar dust is sprinkled liberally throughout the complex.

Blaauw has made a remarkable discovery of the expansion of certain associations. For 17 stars near Zeta Persei, all within 30 parsecs of one another, he found that proper motions and radial velocities indicate a group expansion of 12 kilometers per second (Fig. 50). At this rate the group would have expanded to its present size in 1,300,000 years, a very short interval astronomically speaking; it is obvious that the age of the whole system is probably of that order. Although at first we may be startled at the shortness of the time, we cannot help but be pleased since the derived ages of the hottest O and B stars in this association are probably of the same order; this suggests some sort of explosion, a little more than 1,000,000 years ago, in which the stars were produced and then shot out into space.

Blaauw and Morgan have studied other associations in similar fashion. For a group of about 30 stars near the O star 10 Lacertae they find a rate of expansion of the order of 8 kilometers per second, suggesting a probable age of the order of 4,200,000 years. According to Blaauw, there is evidence that the Scorpius-

49. The Horsehead Nebula in Orion (south of the star Zeta Orionis) photographed in red light with the 200-inch Hale reflector.

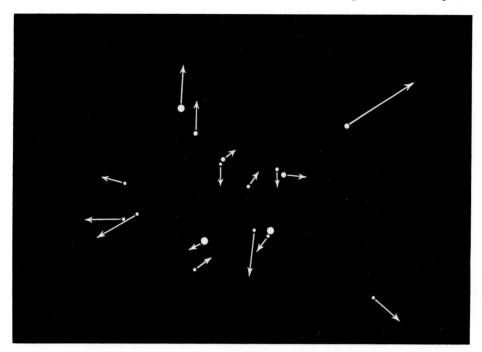

50. An expanding association. The figure shows the Zeta Persei association with the arrows indicating the directions in which the stars are traveling and the distances they will cover during the next 500,000 years. The motions of this group have been studied by Blaauw. (Courtesy of *Scientific American*.)

Centaurus moving cluster is an expanding association, with the slow expansion rate of 0.7 kilometer per second. This gives a probable age of 70,000,000 years, which seems in line with the fact that the Scorpius-Centaurus cluster does not contain excessively luminous, very hot, and hence very young O and B stars, but seems to specialize in the more sedate and older varieties of B stars.

By far the most spectacular expansion phenomenon has been observed by Blaauw and Morgan in connection with the Orion aggregate. The three stars AE Aurigae, Mu Columbae, and 53 Arietis have motions that seem to carry them away from the center of the Orion aggregate at rates in excess of 100 kilometers per second (Fig. 51). They appear to be the hottest and the fastest pellets shot into space by the big Orion explosion, which must have taken place about 2,500,000 years ago.

O and B stars are not the only variety of stars occurring in associations and aggregates. The T associations, discovered and so named by Ambartsumian, contain an abundance of T Tauri-type variable stars. These irregular variable stars are commonly found especially at the edges of very obscured regions. They are probably very young stars of low mass, and they can only be recognized either by their variability in light or by bright-line features in their spectra. They occur in groupings, some of them near O-B associations, others by themselves, but always in regions of the sky where cosmic dust is plentiful. In addition to the Soviet astronomers, Haro at the Tonanzintla Observatory in Mexico and

51. Runaway stars from Orion. The stars AE Aurigae, Mu Columbae, and 53 Arietis are fast-moving
stars. When their paths are retraced they are
found to intersect in the constellation Orion. It is
assumed that they originated in the bright O association and have erupted from it. (Diagram by
Blaauw and Morgan, courtesy of *Scientific American*.)

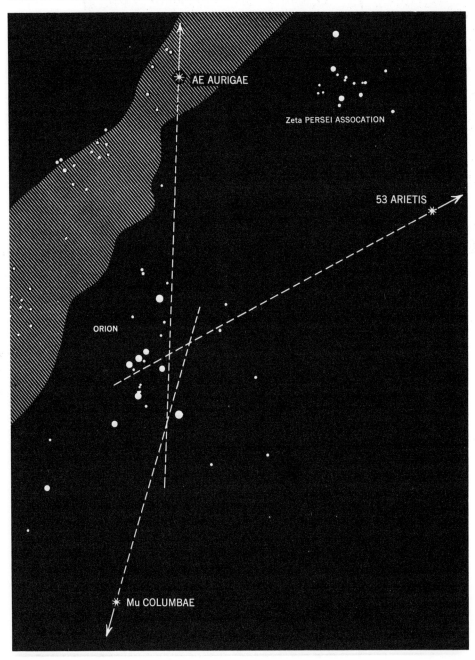

Herbig at the Lick Observatory have searched for and studied the T associations. Holopov in the Soviet Union has catalogued the principal T associations and he was one of the first to show that the T Tauri variable stars are found especially at the edges of dark nebulae.

The Orion Nebula and surroundings form a T association. Apart from the many emission objects discovered in and near the Orion nebulosity, there is an abundance of irregularly varying stars, especially in the regions of dark nebulosity. Originally, many astronomers were inclined to consider seriously the hypothesis that these variations might be caused by variable thickness in the obscuring clouds passing over the stars. Parenago has disproved this hypothesis. The fact that many of the variables have emission lines in their spectra in spite of rather faint absolute magnitudes suggests that the cause of the variability and the emission features lies either in the star's atmosphere or, possibly, in its interaction with the surrounding cosmic dust and the associated interstellar gas. Ambartsumian

is of the opinion that the T Tauri variables are young stars, formed quite recently from the surrounding dust and gas clouds, but that their variability and the presence of emission lines are effects caused by internal disturbances in these youthful stars rather than by interaction phenomena between the stars and the interstellar clouds. Haro and Herbig find support for this suggestion in the presence of excessive ultraviolet radiation in the spectra of these objects. Herbig has some photographs which suggest that we may actually be observing some stars in the process of formation. The variable star FU Orionis is related to the T Tauri variety of variable stars. It came onto the cosmic scene almost unannounced when Herbig found that in 1936 it increased in brightness by 6 to 7 magnitudes. Haro has recently discovered a variable star in Cygnus which in 1969 rose in brightness by at least 6 magnitudes. The T Tauri stars must be evolving very fast. Kuhi has estimated that matter escapes from these stars at a rate of about 1 solar mass in 1.5 million years.

5
Pulsating Stars and Globular Clusters

In the preceding chapters we have gradually moved out from the immediate vicinity of the Sun into other parts of our Galaxy, reaching more remote clusters, some as far as 5,000 parsecs from the Sun and Earth. However, we have covered only a small part of the whole of our Home Galaxy. The study of spectrum-luminosity classes and of precise colors and magnitudes of faint stars has taught us much about the properties of the nearer parts of our Galaxy, but a grand picture of the outline of the Galaxy has not appeared as yet. Our established lamps of known standard brightness just cannot be observed at sufficiently great distances to give us the over-all picture. We must find new standard lamps if we wish to penetrate into the heart of our Galaxy, and beyond. The pulsat-ing stars, especially the long-period cepheid variables and the RR Lyrae stars, provide these lamps. They are the standards that give us the distances to the globular clusters. The outlines of the system of the globular clusters will in turn yield the first definite clues regarding the over-all outline of the whole of our Galaxy.

Pulsating Stars

Among the stars that vary in brightness, some do so because they are changing in size; the star is alternately expanding and contracting. Density waves propagate outward into the lower atmosphere, which expands for a while, then contracts. The star's light varies during the sequence of expansion followed by contraction. The period of light variation

52. Light curve of Delta Cephei. The diagram illustrates the changes in apparent magnitude. The period of the light variation is 5 days 9 hours.

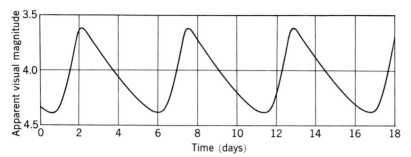

may lie anywhere between 80 minutes and 100 days, but each star has its own characteristic period, which remains constant for most stars within very narrow limits. The brightest star of this class is the star Delta of the constellation Cepheus, from which these stars received the name of *cepheids*. Delta Cephei is an easily identified star; if it is closely observed for a week or two, it is seen to change its brightness between the third and fourth apparent magnitudes in a regular pattern, repeating itself every 5 days 9 hours. Figure 52 shows the light curve; the time is plotted along the horizontal scale and the apparent magnitude along the vertical scale. The curve shows that Delta Cephei rises quickly to its greatest brilliance, fading away more slowly. Over and over again, unvaryingly, it repeats this pattern of changing brightness.

Together with the change in luminosity there occurs a change in color, so that the star becomes redder as it becomes fainter. With the aid of the spectrograph it is shown that the radial velocity of the star varies with the same period as the change in brightness. This variable radial velocity is interpreted as origi-

nating from a periodic swelling and shrinking of the star, a real *pulsation*. The time of greatest velocity of approach comes generally at or near the time of maximum light and the greatest velocity of recession comes at or near the time of minimum. The pulsation theory was first advanced by Shapley to explain the behavior of this star and the other cepheid variables. This theory was developed by Eddington, and later M. Schwarzschild showed that, in addition to the standing waves in the main body of the star, running waves appear near the surface.

A wide range of periods has been observed, but certain periods are favored. Many of these stars have periods of nearly a week or slightly longer. There are, however, fainter stars in the sky that also wink on and off but in shorter periods, of the order of half a day (Fig. 53). The short-period cepheids are frequently called "cluster variables," because Bailey at Harvard first found that they are present in abundance in globular clusters. We shall prefer to call them RR Lyrae stars, after their brightest member.

When we collect the RR Lyrae stars in one group and the pulsating stars of longer periods

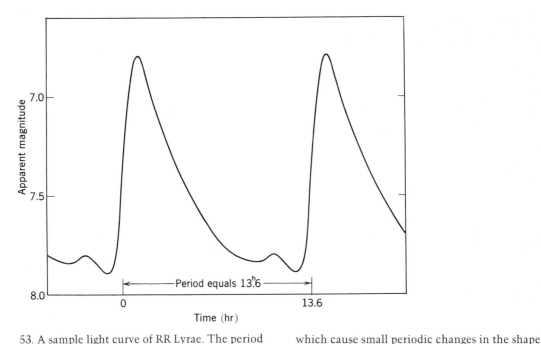

53. A sample light curve of RR Lyrae. The period of this rapidly varying star is 0.567 day = 13.6 hours; its range in brightness is 1.1 magnitudes. The star shows secondary changes in brightness, which cause small periodic changes in the shape of the light curve. (From data assembled by Walraven at the Leiden Observatory.)

in another group, we find that they may be alike in that they are pulsating stars, but they are found to differ in other important respects. The longer-period, or classical, cepheids are mostly at low galactic latitudes and are slow-moving bodies; both their proper motions and their average radial velocities are small. The RR Lyrae stars, on the other hand, move rapidly and are found scattered over the sky; they do not cling to the plane of the Milky Way. Here is a warning that the two varieties of pulsating stars are different sorts of objects, probably different in origin and in age; we must not treat them as members of one single homogeneous group.

The RR Lyrae stars are true Population II stars. They are the sort of stars that inhabit the thin outer halo of the Galaxy and the central nucleus, but they avoid the central disk and the regions of spiral structure. The long-period cepheids that we observe most readily belong to Population I, but here we must be careful. There exist also long-period variable stars with light curves that differ not too much from those of the regular cepheids, but with different mean absolute magnitudes and with characteristics of Population II objects.

During the past 50 years, the pulsating variables have played an important though rather controversial role in the drama of our unfolding knowledge of the outline and dimensions of the universe of galaxies. The first act opened in 1910, when Miss Leavitt of the Harvard Observatory studied the variable stars in the Small Magellanic Cloud. Their periods were found to vary from a few days to over

100 days. She found a very important relation between their periods and their average brightnesses: the longer the period, the brighter the star appears on the average.

Hertzsprung, and later Shapley, recognized this relation as being an intrinsic quality of the stars. Since the thickness of the Small Magellanic Cloud is small compared with its distance, all stars in this star system may be assumed to be essentially equally distant, and a relation between the period and the apparent magnitude will really be one between period and absolute magnitude. The difficulty is to find the constant, or *modulus*, that must be subtracted from all apparent magnitudes in order to change them to absolute magnitudes.

If we only knew the absolute magnitude of a single cepheid of known period, then we could use it to determine the zero point of the period-luminosity curve, and this curve could then be used to give the absolute magnitude of any cepheid for which the period can be observed. Unfortunately, the nearest galactic cepheids are too distant for accurate measurement of their trigonometric parallaxes. Hence it became necessary to determine their average parallaxes from the proper motions by the method of mean parallaxes described in Chapter 3. But the proper motions of these stars are small and their distribution over the sky is not uniform. Both factors introduce uncertainties in the parallax determinations, but it was quite clear that the brightest cepheids are very distant and highly luminous.

The RR Lyrae variables are quite a different story. Since they have high linear velocities, they are especially suited for studies of mean parallaxes. The average radial velocity of a long-period cepheid variable will generally not exceed 20 kilometers per second, but for the RR Lyrae stars velocities of the order of 100 kilometers per second are by no means uncommon. The high linear velocities lead to large proper motions for the RR Lyrae variables of the tenth magnitude and brighter. At the same apparent magnitude, the proper motions of the RR Lyrae variables are far larger than those of the longer-period cepheids. The larger sizes of the proper motions make possible fairly accurate determinations of the mean parallaxes for the RR Lyrae variables.

Originally, the mean absolute magnitudes of the RR Lyrae variables came out close to $M = 0.0$ on both the photographic and the visual scales. Over the years, it has been suggested that this figure might be too high by as much as half a magnitude and mean values for the absolute magnitude between $M_v = +0.5$ and $M_v = +0.7$ are more generally assigned to them. Most astronomers now use the value $M_v = +0.6$ as a suitable mean. The precise value is very important, for, as we shall see below, the assumed mean absolute magnitude of the RR Lyrae variables is the quantity from which the distances to the globular clusters and to the center of our Galaxy are derived.

Until the late 1940's it was assumed that a single period-luminosity curve would include both the RR Lyrae variables and the classical long-period cepheids. The zero point for calibration to absolute magnitude was found by assuming that the absolute magnitude obtained by simply extrapolating the observed curve for periods of the order of a couple of days to periods of the order of 1 day would yield the absolute magnitude of the RR Lyrae variables, that is, $M_v = +0.6$ or slightly fainter. There was some uneasiness expressed about the fact that Shapley and his associates had never been able to discover any RR Lyrae variables in the Magellanic Clouds, but few

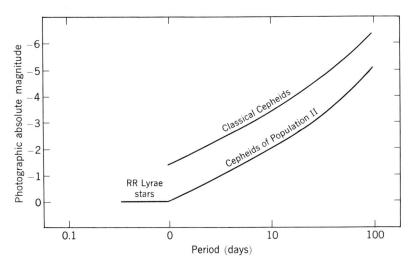

54. The period-luminosity relation. The precise form of the relation is still in doubt, but there is strong evidence to show that the curve for the classical cepheids lies approximately 1.5 magnitudes brighter than that for the Population II cepheids.

astronomers were ready to suggest from this negative observation alone that it is not permissible to extend the period-luminosity curve as indicated. It should be noted that very few astronomers read with care a paper published in 1944 by Mineur in France in which, from a study of motions, he found indications that $M_v = 0.0$ to $+0.3$ probably represents a good mean value for the RR Lyrae variables, but in addition recommended a correction of -1.5 magnitudes to the accepted zero point of the long-period cepheids.

The question became an acute one about 1950, when Baade was unable to detect the RR Lyrae variables in the Andromeda spiral galaxy. Regular long-period cepheids had been discovered in abundance by Hubble 25 years earlier and, on the basis of the accepted period-luminosity relation, the distance to the spiral galaxy had been determined. If the original figure for the distance of the Andromeda galaxy, Messier 31, had been correct, the RR Lyrae variables would have been observed in abundance at about $m = 22$, a

figure close to, but still well above, the brightness limit for the 200-inch Hale reflector. Baade did not find the RR Lyrae variables as expected and he concluded rightly that the zero point of the period-luminosity relation for the long-period cepheids required a correction of just about the amount suggested by Mineur. He recognized, moreover, that the few long-period cepheids observed in globular clusters were of a brand different from the majority of those observed nearby in our Galaxy or in the Andromeda galaxy. The globular-cluster variety of long-period cepheids is Population II (as are the RR Lyrae variables), whereas the regular long-period cepheids, like Delta Cephei and the stars found in the Andromeda galaxy, are Population I. We are now prepared to deal with two period-luminosity relations, one for Population I, another for Population II; the relative shift in zero point for the two parallel curves is of the order of 1.5 magnitudes, the Population I objects being the brighter (Fig. 54).

One concluding comment is in order: a

correction of 1.5 magnitudes is no small matter. It means a doubling of the scales of distance and of diameter for all objects whose distances had been determined earlier with the aid of the older, erroneous curve. In other words, if the correction of 1.5 magnitudes is accepted, then all distances found before 1950 for galaxies outside our own had been underestimated by a factor 2. Our Galaxy escaped a comparable revision of its distance scale only because the RR Lyrae variables were used in establishing its scale, and their assumed absolute magnitudes were not changed appreciably.

Globular Clusters

Whereas there are at most two or three thousand stars in a rich galactic cluster, a globular cluster may contain as many as 100,000 stars (Figs. 55, 56, and 57). There are 120 known globular clusters in the sky and we believe that, unlike the galactic clusters, not more than another 100 lie hidden from us. The globular clusters appear in all galactic latitudes and therefore they are not so often veiled by obscuring matter as are the open clusters; their high intrinsic brightnesses render them conspicuous at great distances. They do not "melt into the landscape" as do the distant open clusters.

The distribution in galactic longitude of the globular clusters is very striking and shouts to all who will listen of the eccentric position of the Sun in our Galaxy. All but a few of the known globular clusters are in one half of the sky and one-third of these are found in a region of Sagittarius that covers only 2 percent of the sky (Fig. 58). The globular clusters form a vast system of their own, concentric with the Milky Way system, but spherical in outline.

Until Shapley's classic investigations of

1916–1919, no estimates of the distances of globular clusters were available. His studies of the RR Lyrae variables discovered by Bailey and others led to the first accurate distance estimates for these far-away objects. With the aid of the 60-inch and 100-inch reflectors of Mount Wilson Observatory, Shapley photographed great numbers of the variables frequently enough to obtain accurate light curves, and, consequently, the distances of all globular clusters with known cluster variables were found.

For the globular clusters that lacked variable stars, estimates of the distances were found from the magnitudes of the brightest stars. Shapley noticed that the brightest stars in the globular clusters—unlike the brightest stars in the neighborhood of the Sun—are not blue-white giants, but are red giants of absolute magnitude about −3, much brighter than the typical red giants found near the Sun.

Rough estimates of distance for the faintest and most distant clusters were made from the observed apparent diameters and apparent integrated magnitudes. Here Shapley was on shaky ground, since he had to make the assumption that all globular clusters are intrinsically identical objects and that apparent differences are due to one variable only—the distance.

At present, cluster-type variables have been identified in about one-third of the globular clusters. Mrs. Hogg has listed 1,100 known variables. For 20 more clusters rough distances have been found from the brightest stars. To find the distance from our Sun to a given globular cluster, we must know not only the apparent magnitude of the RR Lyrae variables in the cluster, but also the amount of intervening space absorption between the Sun and the cluster. This is determined from

55. The globular cluster Messier 92. A photograph of 85 minutes' exposure made with the 61-inch astrometric reflector of the United States Naval Observatory Station at Flagstaff, Arizona. (U. S. Naval Observatory photograph.)

56. The southern globular cluster 47 Tucanae. A photograph of 60 minutes' exposure made with the 40-inch Boller and Chivens reflector of the Siding Spring Observatory of the Australian National University. (Mount Stromlo Observatory photograph.)

57. The southern globular cluster Omega Centauri. A photograph of 60 minutes' exposure made
with the Armagh-Dunsink-Harvard telescope of
Baker-Schmidt design of the Boyden Observatory
in South Africa. Note that this cluster is definitely flattened, suggesting the presence of rotation. (Harvard Observatory photograph.)

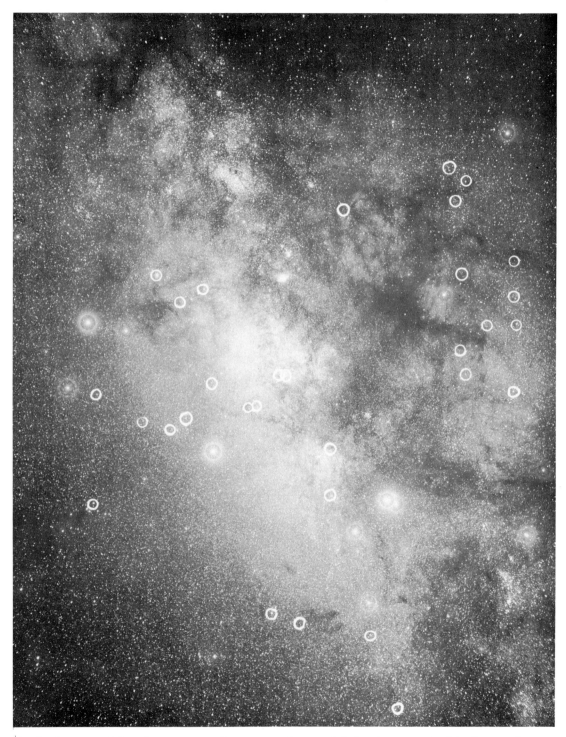

58. The direction of the galactic center. The circles mark the positions of globular clusters for this section of the Milky Way. One-third of all known globular clusters are in this photograph, within an area of only 2 percent of the sky. (Harvard Observatory photograph.)

the amount of space reddening. Mayall has found the integrated spectral types for a number of globular clusters; they vary between A5 and G5, with most of them between F8 and G5. Because of this narrow range of integrated spectrum, the true color index of a globular cluster can be predicted fairly well, and, from the difference between the observed and the intrinsic colors of the cluster, the reddening produced between the Sun and the cluster may be found. By multiplying this color excess by a suitable factor, the total amount of intervening absorption is obtained and the distance can be corrected for absorption effects. Stebbins and Whitford measured the color indices for 68 globular clusters. Their derived total absolute magnitudes for the globular clusters with known RR Lyrae variables range from $M = -5$ to $M = -8$, which is too wide a range to give much confidence in the distances derived from integrated apparent magnitudes alone.

The known system of globular star clusters in our own Galaxy is roughly spherical, with its center located at a distance of 10,000 parsecs from the Sun in the direction of the constellation Sagittarius. We have every reason to suppose that it is concentric with our own galactic system of stars, dust, and nebulae.

Population Characteristics of Globular Clusters

Shapley first pointed out that the stars found in globular clusters differ in several respects from those found in the neighborhood of the Sun. The brightest stars are red and they are some 3 magnitudes brighter than the brightest blue-white stars of the cluster. Shapley was unable to observe stars much fainter than about absolute magnitude $+1$; hence the limit refers to stars that are considerably brighter than our Sun.

With the 200-inch Hale telescope on Mount Palomar it is possible, for the nearest globular clusters, to make extensive studies of colors and magnitudes of stars as faint as our Sun. Thorough studies have been made of Messier 92 by Arp, Baum, and Sandage; of Messier 3 by Sandage and H. L. Johnson; and of Messier 13 by Baum and others. The basic magnitude sequences are generally established by photoelectric measurement. Some of the standard-sequence stars—those of the twenty-third magnitude—are so faint that they cannot be seen visually through the eyepiece of the telescope, though their positions are measurable on photographic plates. They are determined by offsetting for difference of position against a brighter star. For the faintest stars it may be necessary to carry on lengthy comparison measurements for star and background and thus obtain a photometer deflection for the star corrected for background brightness.

After the basic magnitude sequence is established photoelectrically, the magnitudes of other stars may be interpolated by photographic means. Some 1,100 stars were measured in both Messier 3 and Messier 92. Colors can be determined by using blue and yellow filters with the photoelectric photometer and blue- and yellow-sensitive plates with suitable filters for the photographic work.

Assuming the RR Lyrae stars to possess a mean absolute magnitude of $+0.6$, or thereabouts, the distance moduli for the nearer globular clusters are found to be about 14 magnitudes. This means that the difference $m - M$ amounts to $+14$ or $+15$ magnitudes. Hence a star with $m = 20$ will have an absolute magnitude of $+6$, not unlike our Sun; observations to the twentieth magnitude or fainter give a good segment of the main sequence to the limit of stars of the same luminosity as the Sun. Figure 59 shows a typical color-magnitude array for a globular cluster.

59. The observed color-magnitude array for the globular cluster Messier 3. A diagram prepared by Johnson and Sandage. The visual apparent magnitudes V are shown on the vertical scale; color indices B − V are plotted horizontally. The features noted in the text are identified. (Courtesy of *Astrophysical Journal*.)

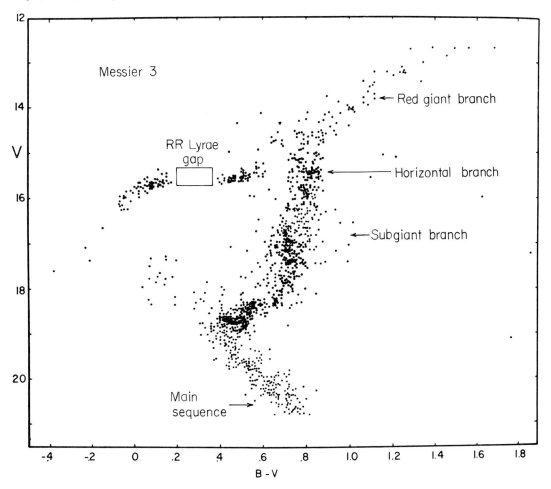

This cluster, Messier 3, also known as NGC 5272, has been studied by H. L. Johnson and Sandage. The diagram shows vertically the visual apparent magnitude V, and horizontally the color index $B - V$. It will be noted that the brightest stars, which are quite red, have apparent visual magnitudes a little brighter than $V = 13$ and that the faintest stars measured by Johnson and Sandage have values of V close to 21. Several features of this diagram should be noted. First of all, we see that between $V = 19$ and $V = 21$ the stars exhibit a well-determined *main sequence.* Whereas in most open clusters the main sequence would continue upward and toward the left, there is a turn-off point at $V = 19$ in the diagram for Messier 3. The stars in the globular cluster have clearly evolved away from the main sequence toward the right, and there is a nearly vertical stretch near $B - V = +0.7$ from $V = 18.5$ to about $V = 15$. We note also two other main features of the diagram for stars brighter than $V = 16$. The first is a scattering of bright stars that outline the sequence toward the upper right-hand corner, the brightest stars, as already noted, having values of V that are a little smaller than 13. All of these bright stars are quite red, their $B - V$ values ranging from 1.2 to 1.7; this is the *red-giant branch.* The stars in the vertical branch near $B - V = 0.7$ are called the subgiants. The second feature is a well-defined *horizontal branch* near $V = 15.7$. It runs from $B - V = -0.1$ to $B - V = +0.6$, but there is a clear gap in this sequence between $B - V = +0.2$ and $B - V = +0.4$. We should mention here that in this particular gap we find in Messier 3 nearly 200 RR Lyrae variable stars! Hence the RR Lyrae variables seem to be a feature of the horizontal branch. At the present time the horizontal-branch stars that seem to be con-

stant in brightness, stars on either side of the RR Lyrae gap, are being studied extensively with the most powerful spectrographs available to astrophysicists. The physics of these particular stars can give us some clues to the reasons why in the middle of the horizontal branch we find a section with lots of RR Lyrae variables. We seem to have evidence of the presence of a critical stage in the nuclear evolution of the interiors of these stars. In all probability the horizontal-branch stars, and their cousins the RR Lyrae variables, are stars that have evolved from the red-giant stage after nearly complete exhaustion of their hydrogen nuclear fuel supplies.

Most globular clusters have color-magnitude arrays not unlike that of Messier 3. The principal differences are in the positions of the turn-off points from the main sequence and the thickness of the subgiant branch, which connects the horizontal branch and the giant branch. The RR Lyrae gap lies sometimes a bit toward the red and in other cases a bit toward the blue from that shown for Messier 3.

There are basically two ways of judging the distance of a globular cluster and of calibrating the vertical scale of apparent visual magnitudes V in terms of visual absolute magnitude M_v. The first is to assume that $M_v = +0.6$ for the RR Lyrae variables; the second is to match the main sequence with the zero-age main sequence as derived from nearby stars and open clusters. Both systems have their weaknesses and points of strength, but, generally speaking, the distance moduli thus arrived at are in fair agreement.

The number of RR Lyrae stars differs very greatly from one globular cluster to the next. Messier 3 has about as many as any of them and there are apparently some clusters that have none. However, wherever RR Lyrae vari-

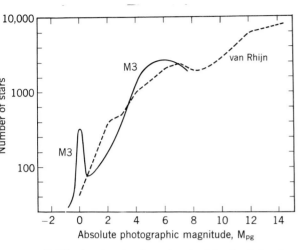

60. The luminosity function for Messier 3. Sandage has counted stars in successive intervals of photographic absolute magnitude in the globular cluster Messier 3. In the diagram his derived luminosity function is compared with that of van Rhijn in Fig. 45, which applies to the vicinity of the Sun. (Courtesy of *Astronomical Journal*.)

globular clusters were made by Shapley about 50 years ago. The subject remained rather neglected until Sandage made a study of the luminosity function of Messier 3. Since his photographic plates reached to apparent magnitude 22.5 and since the distance modulus of this cluster is about 15, he observed stars with absolute magnitudes from -3 to $+7$. From plates of different exposure times, he counted the number of stars of different magnitudes to within 8 minutes of arc from the center. His luminosity function for Messier 3, based on 44,000 stars, is given in Fig. 60, where it is compared with the luminosity function of van Rhijn for the neighborhood of the Sun.

The hump at zero magnitude is very evident. It includes some 200 cluster-type variables and other stars with color indices to the blue of $+0.4$. The drop after $M = +5.5$ Sandage feels is undoubtedly real, though the slope of the curve beyond that point is of course uncertain.

From the number of stars of each absolute magnitude it is possible to compute the contribution of stars of each magnitude to the total light emitted by a cluster. The value from the summation of all contributions is $M = -8.09$, which is in good agreement with the integrated absolute magnitude found by Christie. Ninety percent of the light of the cluster comes from stars brighter than the fourth absolute magnitude, that is, brighter than the Sun, although the total number of stars like the Sun and fainter is very large.

Since it is impossible at present to go below $M = +7$, Sandage uses the van Rhijn luminosity function to estimate the total number of stars of all luminosities in the cluster and from these numbers he can estimate the cluster's total mass, which he finds to be 1.4×10^5 solar masses. Wilson and Miss

ables are found, there are apparently no stars of constant brightness at that particular position in the color-magnitude array. Hence it follows that the RR Lyrae gap is truly an evolutionary stage of instability in the globular-cluster stars.

Globular clusters are the oldest known objects related to our Galaxy. The turn-off point from the main sequence is indicative of the probable age of the cluster. The ages of globular clusters range from 5×10^9 years to 2×10^{10} years. The globular clusters truly all have the appearance of being old and worn-out stellar systems. Neither optically nor by radio-astronomical techniques does one find evidence of appreciable amounts of free gas or dust associated with globular clusters. In them the processes of star formation have apparently ceased to operate long ago.

Early studies of the luminosity functions in

61. Distribution of globular clusters. The diagram
shows the distribution of globular clusters with
known distances projected on a plane passing
through the Sun and the galactic center perpendicular to the central galactic plane. The position
of the Sun and of the galactic center are indicated. The vertical scale represents the height z of
the cluster above ($+$) or below ($-$) the galactic
plane measured in kiloparsecs (1 kpc = 1,000
parsecs). The horizontal scale x is measured in the
same units. (Diagram prepared by Arp, reproduced in *Galactic Structure*, courtesy of University of Chicago Press.)

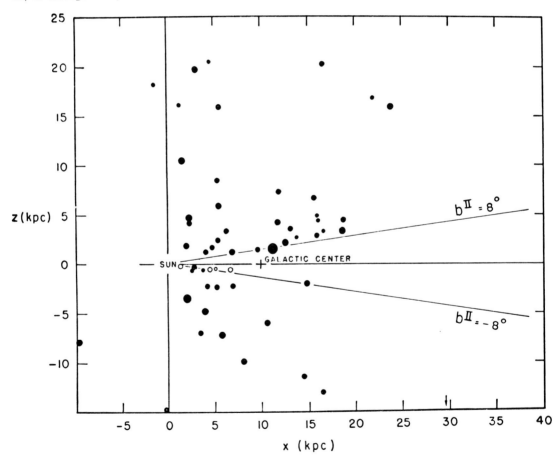

Coffeen estimated the total mass of Messier 92 from dynamical theory as 3.3×10^5. In view of the greater luminosity of Messier 3 as compared with Messier 92, its dynamical mass should be of the order of 5.6×10^5, larger than the above value by a factor 4. It is still uncertain whether this discrepancy is real. If the dynamical value for the mass is correct, then the result may indicate that the percentage of faint stars is much greater in a globular cluster than in the neighborhood of our Sun.

The System of Globular Clusters

In spite of uncertainties in the mean absolute magnitudes assigned to RR Lyrae stars, these stars still provide the best standard lamps for fixing the distances to globular clusters. We now have available good photoelectric photometry to the limit of the brightness of the horizontal branch for close to 25 globular clusters, and for these one can obtain reasonably reliable distances, which, incidentally, can be corrected for the effects of interstellar absorption. A few years ago Arp put together all of these distance estimates and prepared a list of distances for 21 globular clusters. Messier 3, the cluster to which we referred at some length in the preceding section, is among those and Arp gives it a true distance modulus of 15.13, to which corresponds a distance of 10,600 parsecs. For clusters without reliable variable-star photometry, approximate distances can be derived on the basis of the mean absolute magnitude for the 25 brightest stars. Shapley used this criterion

in his investigations of half a century ago. Arp assigns to the mean absolute magnitude of the 25 brightest stars a value $M_v = -0.8$. This helps in fixing the distances for quite a few more globular clusters. Other criteria, such as integrated brightness and apparent diameter, have been used in the past, but these have been shown to be too unreliable for good work.

Figure 61 is Arp's diagram of the distribution in our Galaxy of the globular clusters of known distance. The diagram represents a cut in the plane passing through the Sun and the galactic center at right angles to the central galactic plane. The positions of the Sun and of the galactic center are marked in the diagram. The Sun is probably 10,000 parsecs distant from the galactic center. The tilted lines drawn in the diagram starting at the Sun make a section in which most globular clusters will be hidden from our view by interstellar obscuration close to the central plane. The globular clusters can be divided into two groups, one fairly well concentrated to the center of our Galaxy, the other far from the galactic plane. We note that studies of integrated spectra of globular clusters for the two groups show that there seem to be rather marked differences in chemical composition between the central group and the outlying clusters. The latter seem to be generally metal-poor in their spectra, whereas metallic lines do show up more prominently in the spectra of the clusters found close to the center of our Galaxy.

6
The Whirling Galaxy

Thus far we have dealt only with the purely structural aspects of the Milky Way problem. Now we shall turn to considerations involving the motions of the stars and the forces that control them; we shall see what may be learned about the *dynamics* of the system from studies of the regularities in the proper motions and radial velocities of the stars and of clusters of stars.

We found the Milky Way system to be a highly flattened disk, embedded in a very tenuous, more or less spherical, halo. The characteristic spiral arms, outlined by the O and B stars and by bright and dark nebulae, lie in or very near to the central plane of the Galaxy. At the center of the disk there is a very dense nucleus, apparently composed mostly of older stars. The Sun is probably about 10,000 parsecs from this dense central nucleus, but it is within 30 parsecs of the central plane.

It should have been evident all along that the flattened shape of our Galaxy is indica-tive of rapid symmetric rotation about an axis perpendicular to the plane, but it was not until 1926–27 that the researches of Lindblad and Oort first gave conclusive proof of this *galactic rotation*. In the first section of this chapter we shall present the evidence to show that our Sun and the great majority of the stars in our vicinity move about the center in roughly circular orbits and with speeds of the order of 250 kilometers per second. Most of the regularities observed in the motions of the stars near the Sun and those at distances up to a few thousand parsecs can be understood on the basis of this galactic rotation.

The Sun's Motion around the Galactic Center

The highly flattened shape of the system of O and B stars, and even of the system of stars like our Sun, suggests that these subsystems of our Galaxy are in rapid rotation. The situation is quite different for the globular clus-

62. Spectrum of the globular cluster Messier 15. This photograph shows the spectrum of the cluster in the middle, with a comparison spectrum (see Fig. 39) above and below the spectrum of the cluster. Rarely are the absorption lines as sharp as those shown here. (Photograph by Kinman, courtesy of Lick Observatory.)

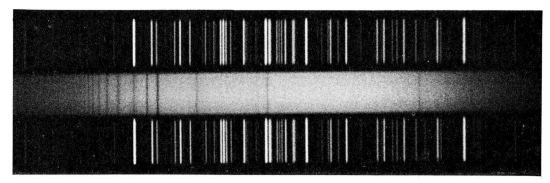

ters, which form a tenuous and more nearly spherical subsystem surrounding the flattened principal Milky Way system. Inside a system of nearly spherical shape there may be large random motions, but there is probably no rotation of the system as a whole. On the average, the globular clusters should therefore define a system more or less at rest with respect to the center of our Galaxy. From the study of the observed radial velocities of globular clusters, we should be able to obtain information about the Sun's motion relative to the system of globular clusters and hence presumably with reference to the center of our Galaxy.

Observed radial velocities are available for 70 globular clusters. The majority of these radial velocities were measured by Mayall at the Lick Observatory; his basic list contains 50 values. Radial velocities for 20 additional globular clusters, mostly at southern declinations, were obtained by Kinman at the Radcliffe Observatory in South Africa. Mayall found individual values ranging between

+291 and −360 kilometers per second. Kinman's values do not cover quite so wide a range. The spectral lines are broad and fuzzy, and precise measurement of the radial velocities is far from simple (Fig. 62). If we assume that the globular cluster system as a whole has at best only a small residual circular velocity of rotation with respect to the center of our Galaxy, then we should be able to obtain from the observed radial velocities of the globular clusters the average velocity of the Sun with respect to the system of globular clusters. Such a study would presumably yield a lower limit to the circular rotational velocity of the Sun around the galactic center. For an accurate determination of the motion of the Sun relative to the system of globular clusters, we require a fair number of clusters in directions of galactic longitude $l = 90°$ and $l = 270°$; the radial velocities of these clusters would show the full effects of the Sun's motion (directed more or less toward galactic longitude $l = 90°$). Unfortunately, as we see from Fig. 61, the globular clusters, when viewed from the Sun, are concentrated to one hemisphere, and the majority of them are seen in directions from the Sun that do not differ much from the general direction to the galactic center, galactic longitude $l = 0°$.

In spite of these obstacles, Mayall and Kinman were able to derive values for the Sun's velocity relative to the system of globular clusters. The maximum value of this circular velocity, which is directed to a point in the galactic plane at galactic longitude close to 90°, is 200 kilometers per second. Mayall's values are a bit higher than Kinman's. However, as Mayall pointed out many years ago, the value of 200 kilometers per second for the rotational velocity of our Sun relative to the galactic center is a minimum value, since it is not unlikely that the system of the globular clusters—which shows some slight degree of flattening—may have a residual rotation of the order of 50 kilometers per second. In summary, the Sun's motion relative to the system of globular clusters is probably of the order of 200 kilometers per second.

Evidence to show that the circular velocity of the Sun and its neighbors is greater than 200 kilometers per second has come from radial-velocity data for neighboring galaxies determined mostly at the Hale Observatories and at Lick Observatory. Together with the Large and the Small Magellanic Clouds, the spiral galaxies in Andromeda and Triangulum, and about a dozen smaller objects, our Galaxy forms a sort of local supersystem. If we assume a velocity of the order of 250 to 270 kilometers per second toward galactic longitude 90° for the Sun relative to the center of our Galaxy, we find that the average relative speeds of all galaxies that belong to our local system of galaxies are less than 50 kilometers per second.

Following the lead of the International Astronomical Union, Milky Way astronomers have in recent years come to assume a value of 250 kilometers per second as the most likely one for the circular motion of the Sun relative to the center of our Galaxy. This value fits nicely into the over-all scheme for the constants of our Galaxy.

It may seem from the arguments presented in this chapter and the preceding one that astronomers are reasonably well in agreement on the values assumed for two basic constants of our Galaxy: 10,000 parsecs for the distance from the Sun to the galactic center and 250 kilometers per second for the circular velocity of the Sun relative to the center of the Galaxy. This is not the case. There is much discussion among the experts relating to both "constants." In recent years the generally assumed value, $M_v = +0.6$, for the mean absolute magnitude of the RR Lyrae variables has been repeatedly questioned. Clube especially has come forth with evidence to show that the mean absolute magnitude for these stars may be fainter by one-half to as much as a full magnitude. However, work by Graham on RR Lyrae variables in the Large Magellanic Cloud supports the mean value $M_v = +0.6$, and shows that the spread around the mean is small, of the order of ±0.2 magnitude. Clube's revised value of 7,000 parsecs for the distance from the Sun to the galactic center has so far not seemed acceptable to the majority of galactic astronomers. However, a 15-percent reduction (from 10,000 to 8,500 parsecs) does not seem to be out of the question. It is well that our readers should at all times bear in mind that most quoted distances in our Milky Way system are uncertain by ±10 percent, and there are even greater percentage uncertainties in the extragalactic distance scale. We end on a cheerful note: no one has yet questioned that our Sun moves in a very nearly circular orbit around the galactic center.

The Stars of High Velocity

Most stars near the Sun move with speeds relative to the Sun not in excess of 30 kilometers per second, but there are some that exceed this limit by a wide margin. We are accustomed to refer to the stars with velocities in excess of 60 kilometers per second relative to the Sun as stars of high velocity. Miss Roman has published a catalogue of 600 such stars. In Fig. 63 we reproduce a diagram in which the velocity arrows for the high-velocity stars within 20 parsecs of the Sun are shown projected on the plane of the Milky Way system. With practically no exception, the stars are seen to move toward the half of the Milky Way between galactic longitudes $l = 180°$ and $l = 360°$ (from Auriga, through Orion, to Carina and Sagittarius-Scorpius). Not a single star of high velocity is seen to move in a direction toward $l = 90°$, which is in the general direction of the Cygnus section of the Milky Way. If we determine the average motion of the stars shown in Fig. 63 relative to our Sun, we find it directed toward galactic longitude $l = 270°$. The apex of the Sun's motion relative to the high-velocity stars agrees with that derived from the radial velocities of the globular clusters.

The RR Lyrae variables are the best-known group of high-velocity stars; they give a value for the Sun's apex close to $l = 90°$, with an average speed relative to the Sun of 130 kilometers per second. These stars are arranged in a subsystem that is not quite spherical but has a slight flattening; the RR Lyrae stars are highly concentrated in their distribution toward the nucleus of our Galaxy. The observed flattening of the system of the RR

Lyrae stars suggests some residual rotation of the system as a whole, though considerably less than for our Sun; the observed degree of flattening is consistent with a rotational speed of $250 - 130 = 120$ kilometers per second for the RR Lyrae stars as a whole relative to the center of the Galaxy. The purely random motions of the RR Lyrae stars (found after correction for systematic motions) come out quite high—of the order of 100 kilometers per second.

The amazing characteristics of the motions of the high-velocity stars were first studied thoroughly in the years 1924 to 1926 by Oort and by Strömberg. One of the most extensive surveys of this variety of stars was published in 1940 by Miczaika, who at that time was able to list 555 stars with velocities in excess of 63 kilometers per second. Certain varieties of stars contain much higher percentages of high-velocity stars than others. Practically none are found among the B stars, whereas they occur rather frequently among the M stars, where we find that 19 percent of those fainter than $V = 7$ are high-velocity stars.

Bertil Lindblad, who was the originator of the theory of galactic rotation, developed this theory especially to explain the observed asymmetry in the motions of the high-velocity stars. He thought of the Milky Way system as made up of a number of concentric subsystems, each with its own peculiar degree of flattening and rotational speed relative to the galactic center. Even today this approach retains considerable validity and it has been used extensively in the early work of the Soviet astronomers Kukarkin and Parenago. According to this concept, the Sun, and all the stars with observed low velocities (less than 20 or 30 kilometers per second) relative to it, whiz around the galactic center at a rate close to 250 kilometers per second. The RR Lyrae

63. The asymmetry in stellar motions. The diagram gives the distribution in the galactic plane of the directions of the velocities of nearby stars with speeds in excess of 60 kilometers per second. The galactic longitudes of these directions are indicated. Not a single star in the diagram is found to move in a direction between galactic longitudes $l = 30°$ and $l = 150°$; the center of the sector of avoidance is at $l = 90°$.

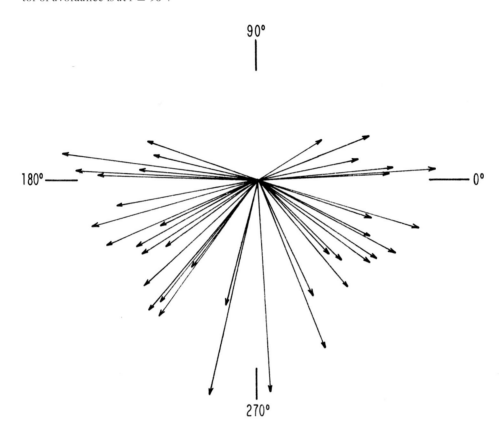

variables, moving with an average speed of 130 kilometers per second relative to the Sun in the direction opposite to where the Sun is heading, are really the laggards, and their average rotational speed is no more than $250 - 130 = 120$ kilometers per second relative to the center of the Galaxy (Fig. 64).

But, you might ask, why are there not also some stars of really high velocity moving with speeds of 100 kilometers per second relative to the Sun in the direction of galactic longitude $l = 90°$? The reason is that such stars would have velocities relative to the center of our Galaxy of the order of 350 kilometers per second, and the Galaxy apparently cannot retain such speeders. A velocity in excess of 310 to 320 kilometers per second relative to the galactic center is apparently sufficiently great to permit a star with such a velocity to escape from the Galaxy, or at least to place it in a rare outlying orbit.

Solar Motion and Star Streaming

We return now briefly to a consideration of the motions of the normal stars of all spectral classes and luminosities that are relatively near to the Sun. We saw in Chapter 3 that the Sun moves with a speed of 18 to 20 kilometers per second toward a solar apex in the constellation Hercules. This result is obtained when the motions are referred to the average for the nearby stars as a standard of rest.

The velocity of the Sun is found not to have the same value for all classes of stars. As early as 1910, Campbell at the Lick Observatory suggested that the derived value of the solar velocity varies with spectral class. He found that the B stars give a rather small solar velocity and that its value increases as we proceed from spectral class A through F to G and K but that it becomes somewhat more like the B average for the bright M stars. Later

work has shown that there is a marked dependence of the value of the solar velocity upon the intrinsic luminosities of the stars, with the M dwarfs, for example, giving a much higher solar velocity than the M giants.

The researches of the Vyssotskys at the University of Virginia showed that both the position of the apex and the value of the solar velocity depend markedly on the spectral class and on the average apparent magnitude of the stars with respect to which the motion is determined. The position of the apex for the A and F stars is found at lower galactic latitude than for the other types. The differences in position of the apex and in the value of the Sun's velocity are probably due in part to effects of moving clusters, but they are caused mostly by an uneven admixture of high-velocity stars and by local streamings, different from pure galactic rotation.

We turn now to the subject of *star streaming*. Throughout the nineteenth century, astronomers were blissfully unaware of any regularities in stellar motions beyond those arising from the reflex of the Sun's motion. Then, in 1904, Kapteyn of Holland announced the discovery of two star streams. With this discovery came the first realization that the stars are not moving in a perfectly haphazard fashion, but that their motions are subject to general laws.

Kapteyn's researches dealt with the proper motions of the brighter stars in the sky. The celestial sphere was marked off into a number of sections, in each of which Kapteyn counted the number of stars moving within certain narrow limits of direction. A plot of the proper motions on a chart or on a celestial sphere showed him directly how many stars were moving within 15° of the direction of the North Pole in the sky, how many within 15° of the northwest, and so on. A diagram

64. Asymmetry and the rotation of the Galaxy. The Sun and the majority of the stars near us move around the center of our Galaxy in roughly circular orbits with velocities of the order of 250 kilometers per second. A star that would move in an elongated galactic orbit such as is shown here might have a velocity of only 120 kilometers per second at the point where the orbit reaches as far as the Sun. This star would be observed from the Sun and Earth as having a "high" velocity of 130 kilometers per second directed opposite to that of the general galactic rotation. If it were observed to have a velocity with respect to the Sun and its neighbors as high as 100 kilometers per second in the direction of galactic rotation, then its speed relative to the galactic center would be 350 kilometers per second and the star would in all likelihood not be a permanent member of our galactic system.

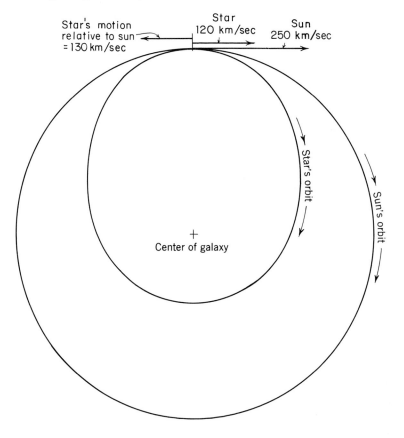

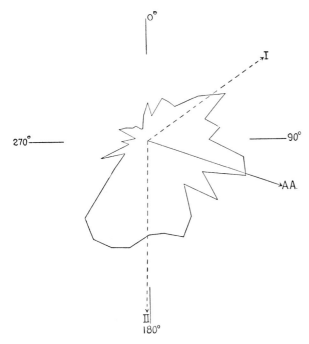

65. A typical star-streaming diagram. The diagram illustrates the distribution of the directions of proper motions, measured by Smart, for a small region of the sky; the arrow *AA* marks the direction toward the antapex. The broken arrows I and II show the directions of the two star streams for this section of the sky.

such as Fig. 65 summarizes the observed distribution in a convenient fashion. For a given section of the sky, we draw for each particular direction an arrow, the length of which is proportional to the number of stars found traveling in that direction. By connecting the ends of these arrows with straight lines, we obtain a clear representation of the distribution of the directions of proper motions for that section of the sky.

If the stars were moving perfectly at random, and if the Sun had no motion of its own, the lines connecting the points in the diagram would form a circle. The Sun's mo-

tion would have the effect of drawing this circle out into an elongated figure resembling an ellipse, the long axis of which should point away from the apex of the Sun's motion. For fields not far from the apex and antapex the figure should be almost circular, and the greatest flattening should be found along the circle on the sphere halfway between the apex and the antapex. In fact, however, we find nothing of the kind. The curves near the apex and the antapex are by no means circular, and simple figures shaped like ellipses are not observed. The characteristic figures are bilobed. Generally there are two directions that the stars in a given section seem to prefer; these are marked I and II in Fig. 65.

It was soon found that these streaming effects are not purely local ones. For each section we can draw the two preferred directions, and, if we plot these directions on a celestial sphere, the great circles defined by them converge to two points on the sphere. These points were called by Kapteyn the *apparent vertices* of his two star streams. If the observed streaming tendencies had been of a random character, we would presumably have found no regularity in the distribution of these arrows over the sphere. The fact that each stream shows a well-marked convergent point is excellent proof that the stars all over the sky show in their motions preference for either Stream I or Stream II.

The Kapteyn star streaming bears some resemblance to the convergence effect observed for the motions of stars belonging to a moving cluster, such as the Hyades cluster. In the case of the moving cluster the proper-motion vectors of all individual members are found to pass exactly through the convergent. The motions of the stars that belong to one of Kapteyn's streams show only a tendency to move along with the general stream motion rather than at right angles to it. They still insist on preserving their right to deviate considerably from the path of streaming instead of submitting to a Hyades-like regimentation.

It is not difficult to correct for the effect of the Sun's motion and derive the position of the vertices as viewed from a star supposed to be at rest with respect to the average of all of its neighbors. Kapteyn showed that the *true vertices* of star streaming, found after correcting for solar motion, fall at opposite points in the sky, one in Scutum, the other in Orion.

It was not surprising that the true vertices are 180° apart in the sky, because, by correcting for local solar motion, we have automatically balanced the motions of the two streams. But it was surprising, and highly significant, that the line of the true vertices falls exactly in the plane of the Milky Way, and we note in passing that one vertex has a direction in the sky not far from that to the galactic center. An explanation was not directly forthcoming, but at the time of Kapteyn's discovery it was realized that this represented some new major clue to the ultimate solution of the riddle of the Milky Way system.

Subsequent researches by Eddington and by Karl Schwarzschild, and more recent studies by many other astronomers, most notably the Vyssotskys at the McCormick Observatory, have confirmed and extended Kapteyn's work. Schwarzschild showed that it was unnecessary to think in terms of two specific streams of stars. He pointed out that the line of the true vertices marks the direction along which the stars prefer to move. He found methods for expressing the spreads, or dispersions, of the velocities along the direction of the true vertices and in two mutually perpendicular directions at right angles to the line of the true vertices; the spread is slightly less

(0.6 to 0.7 of the first quantity) for the direction at right angles to this line in the galactic plane, and somewhat smaller still for the direction at right angles to the galactic plane.

Lindblad was the first astronomer to prove that star streaming according to the picture of Schwarzschild is a natural consequence of the rotation of our Galaxy. The majority of the stars move in orbits that are almost, but not quite, circular around the center of the Galaxy. These slightly oval orbits in the galactic plane—which are generally also slightly inclined with respect to the plane—produce observable deviations from pure circular motion, and Lindblad was able to show that, on the average, the stars can be expected to exhibit somewhat greater spreads in their motions toward or away from the center than at right angles to this direction.

According to the purest form of the theory of our rotating Galaxy, the line of the true vertices should pass precisely through the galactic center. This happy state of affairs does not exist, for the vertex for some groups of stars lies as much as 15° away from the direction to the center. The explanation of this observed deviation is probably found in the nonuniform distribution of the stars near the galactic plane. Our Galaxy has spiral arms, and we may expect a rather low density of stars and gas between them. In the arms themselves there are concentrations of many sorts, from extended star clouds and aggregates to expanding associations. We live in a part of the Galaxy that is far removed from the well-mixed state envisioned by the simplest form of the theory of a rotating galaxy. Vertex deviation may well provide the clue to the intricate relations that must exist between local variations in stellar distribution and local velocity characteristics.

Galactic Rotation

After Lindblad had presented his explanation of the observed asymmetry in stellar motions, Oort found a further proof of the general rotation of our Galaxy in the radial velocities and proper motions of the stars between 300 and 3,000 parsecs from the Sun. Oort reasoned that it is very unlikely that our Galaxy would rotate like a solid wheel. If it did, the stars would keep their same relative positions and there would be no evidence of motion except from galaxies beyond our own. Since it appears that a considerable part of the total mass of our Galaxy is concentrated near the galactic nucleus, we may expect that the motions of the stars in our Galaxy will resemble those of the planets around the Sun. Venus moves faster than the Earth and the Earth in turn outruns Mars, and in the same fashion the stars nearer the galactic center could be expected to complete a galactic circuit in shorter times than those farther from the center. We might offhand doubt whether this conclusion would hold for our Galaxy, since (see Table 4) the circular velocity of rotation around the galactic center does not vary quite in the manner of the planets of our solar system. Table 4 does show, however, that the stars closest to the galactic center complete a circuit more quickly than the more distant ones. Take, for example, the value of the circular velocity of an object at 20,000 parsecs from the galactic center. If our Galaxy were rotating like a solid wheel, the circular velocity at 20,000 parsecs would be twice as large—500 kilometers per second—as the velocity of rotation near the Sun; the value listed in Table 4 for 20,000 parsecs is only 193 kilometers per second. The more distant parts of our Galaxy obviously lag behind the nearer ones in the completion of

Table 4. The Schmidt model of circular velocities in the Galaxy.

Distance from center (parsec)	Circular velocity (km/sec)
1,000	200
2,500	190
5,000	227
7,500	250
10,000	250
12,500	235
15,000	218
17,500	205
20,000	193

one full revolution around the galactic center.

Since all our observations are made from the Earth, which accompanies our Sun, we ask what observable effects in the radial velocities there might be for an observer at the Sun. Oort showed that, as viewed from the Sun, some distant stars would seem to be catching up with us, others would seem to be receding. He found that the effect in the radial velocities would go through its range of values twice as we observe completely around the galactic circle. Figure 66 shows how this comes about. In the diagram on the left, the arrows indicate schematically how the velocity of rotation decreases with increasing distance from the galactic center. In the second diagram, showing the region around the Sun, the effect of the solar motion has been removed and the arrows represent the velocities as seen from the Sun. They show the effects of the differential galactic rotation. The components of these velocities along the radii passing through the Sun will be the effects of differential galactic rotation on radial velocities.

66. The effects of galactic rotation on radial velocities. The diagram on the left illustrates a possible variation of rotational velocity with distance from the galactic center. In the diagram on the right we reproduce on a larger scale the region

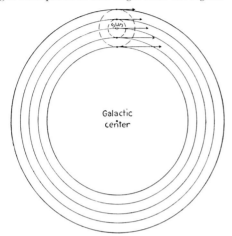

around the Sun. The arrows now represent the velocities as observed from the Sun. The radial components of these velocities are shown and these exhibit the variation of galactic rotation with galactic longitude, as described in the text.

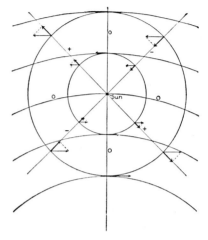

At four points along the circle there should be no approach or recession of the stars due to galactic rotation, and the observed effect in the radial velocities is there zero. These are the directions toward and away from the galactic center and at right angles to that line. Halfway between these points, one diagonal line shows the direction in which the stars will on the average be receding, and hence have positive velocities, and along the other diagonal they should have on the average negative radial velocities, corresponding to approach. As we plot the average observed radial velocities against the galactic longitude, the resulting curve should be a double sine wave, going through two maxima and two minima for the complete circle of longitudes. It is important to note that the galactic-rotation effect in the radial velocities observed at the Sun will be greatest for the most distant stars. Oort showed that for distances up to 2,000 parsecs from our Sun the effect

increases very nearly proportionally to the distance.

It may be of interest to reproduce the simple mathematical expression for the galactic rotation effect in radial velocities in the form given by Oort. If V is the effect in the radial velocity, r the average distance of the stars under consideration, l the galactic longitude of the star, then $V = rA \sin 2l$. This formula may be readily checked against Fig. 66. The factor A in the formula is generally known as "Oort's constant" and it measures the maximum effect in V at the standard distance. The accepted value of Oort's constant A is 15 kilometers per second per 1,000 parsecs. The stars that have proved most useful in studies of the galactic-rotation effect are distant, intrinsically luminous objects like O and B stars, open clusters, emission nebulae, and long-period cepheids.

In the past decade a new approach to the study of galactic rotation has become possible. We can now fix with reasonable confidence an estimated distance for any single O or B star we observe. This is done by combining the absolute magnitude of the star, as found from spectrum-luminosity classification, with its observed apparent magnitude and color index. If we measure the star's radial velocity, then we have available two basic bits of information about the star: its radial velocity and its distance from the sun. A small correction is applied to the star's radial velocity (for the standard solar velocity of 20 kilometers per second) to find its value relative to the standard near the Sun moving in a circular orbit around the galactic center at the standard rate of 250 kilometers per second. Assuming that the distant O or B star moves in a nearly circular orbit around the galactic center, its observed radial velocity will yield a value for its circular velocity around the galactic center. The method that we have described had its origin in work by Camm in the 1930's, but its application has become possible only in recent years with the advent of modern techniques of spectrum-luminosity classification. Improved values of the circular velocities listed in Table 4 should become available in the years to come as distances and radial velocities are determined with precision for increasing numbers of distant O and B stars (and for cepheid variables as well).

Oort's theory of galactic rotation predicts also an effect in the proper motions of the stars. The effect in these transverse motions reaches its extreme values for directions in which the effects of radial velocity are zero. We measure proper motions not in kilometers per second but in seconds of arc per year. If, for a certain direction in the galactic plane, we go twice as far out, the linear effect will be doubled, but in the observed angular motions the effect will remain the same as that measured for the nearer stars. To the distance of 2,000 parsecs the effect in the proper motions should vary only with the galactic longitude of the stars involved and not with their distances.

Unfortunately, it is very difficult to detect and measure the galactic-rotation effects in the proper motions. The quantities to be derived from an analysis of the proper motions are small, of the order of a few thousandths of a second of arc per year. Their signs and precise values are made uncertain by possible systematic errors in the basic system of proper motions and we do not know the constants of precession and nutation sufficiently well for dependable analysis of most large bodies of proper-motion data. Clube has recently analyzed the first proper-motion results obtained from the Lick Survey, in which the motions are referred to a fixed system of faint galaxies (see Chapter 2). The Clube results seem to lend support to his low value (7,000 parsecs) for the distance from the Sun to the galactic center. In addition, Clube finds evidence that the Sun and the stars near it show a systematic motion of the order of 35 kilometers per second (or greater!) away from the galactic center. If this suggestion is supported by further evidence, the result will have far-reaching consequences for the study of the dynamics of our Galaxy.

Before we leave the subject of galactic rotation, we should inquire briefly into the shapes of the orbits in which the stars move in our Galaxy. The Sun's motion deviates only slightly from the circular galactic rotation in its neighborhood. We can think of the Sun as moving in an ellipse of small eccentricity, while at the same time it oscillates slowly back and forth perpendicularly to the galactic plane. The Sun will probably stay at all times within 200 parsecs of the galactic plane, and it should complete between two and three oscillations in the perpendicular direction in the 250,000,000 years that it takes to complete one revolution around the center of the Galaxy.

The Sun and the majority of the Population I stars have motions that differ by not more than 20 kilometers per second, that is, by less than 8 percent, from pure circular motion. Hence all of their orbits differ only slightly from pure circular orbits and these stars are in all likelihood now at more or less the same distance from the center of the Galaxy as they were at the time of their birth. The O and B stars, the cepheids, certain open clusters, and the interstellar gas clouds follow most nearly the circular orbits of pure galactic rotation. Some of these stars—notably the O and B stars—are presumably quite young and we shall see later that their probable ages are only a small fraction of the time of one galactic revolution.

The RR Lyrae stars and other fast-moving stars are presumably following elongated orbits shaped somewhat like ellipses. One wonders how close such stars may once have come to our galactic center. Martin Schwarzschild of Princeton has shown that these stars, in spite of their large peculiar motions, have probably never been much nearer to the center than half-way to the Sun. Miss Roman, then at Yerkes Observatory, has, however, discovered some weak-line F stars with very elongated galactic orbits. Some of these stars have velocities that differ so much from pure circular motion that they must have come from points within 2,000 parsecs of the center of the Galaxy. They are probably the best-known representatives of stars that were shot out by the central region of the Galaxy to our remote outpost; they must have come directly from the heart of the Galaxy. By the study of its presently observed motion, we may obviously learn much about the past history of a star.

The theory of galactic rotation represents a great advance in our understanding of the observed regularities in stellar motions. It has provided us with very reasonable explanations of star streaming and of high-velocity stars, and has further led to the discovery of the Oort effects. It is proving extremely useful in interpreting the observations of the 21-centimeter radiation from neutral hydrogen (Chapters 8 and 10).

We should, however, bear in mind that the picture of a smoothly rotating galaxy can at best be only a first and very rough approximation to the true state of affairs in our complex galactic system. The presence of spiral arms, of expanding clusters and streams, of irregularities of distribution and motions, all contribute to the intricacies of the problem. At the same time, these phenomena may hold the key to the past and to the future development of our Galaxy.

7
The Nucleus
of Our
Galaxy

The Nucleus

Before the work of Shapley (1916–1919), it had been generally assumed that our Sun is not far from the center of our Galaxy and that the stars thin out in all directions from the Sun. Indications were that the diameter of the whole system could hardly exceed 10,000 parsecs. To Shapley belongs the ever-lasting credit of having shown that the Sun and Earth are nowhere near the center of the Galaxy. He gave conclusive evidence of the presence of a massive galactic nucleus in the general direction of the Sagittarius star clouds. Modern estimates place this nucleus at a distance of 10,000 parsecs from the Sun. The apparent thinning out of the stars in all directions from the Sun was explained later by the presence of an interstellar absorbing medium, which dims the light of the distant stars and makes them seem farther away than they really are. Our Galaxy stretches to a distance of at least 5,000 parsecs beyond the Sun, thus yielding an over-all diameter for the central galactic disk of $2 \times 15,000$ parsecs $= 30,000$ parsecs. It is generally estimated that the total mass of the Galaxy is a little under 200 billion (2×10^{11}) solar masses, the precise value, according to the model of M. Schmidt, being 1.8×10^{11} solar masses. About 5 percent of this total mass is concentrated in the nuclear region of the Galaxy, and the remainder appears to be divided rather evenly between the flattened galactic disk interior to the Sun and the outer spheroidal shell of the Galaxy. The nucleus of the Galaxy is its gravitational center, which controls the motions of the stars in our vicinity. The high concentration of matter to the galactic center is responsible for its over-all rotational properties (see Table 4). It is clearly important that we should learn as much as possible about the nuclear regions of the Galaxy.

When the third edition of this book appeared in 1957, we devoted only a few paragraphs to the nucleus of the Galaxy. It had become evident that the center itself was hidden from view by dense overlying obscuration. Optical studies of peripheral objects, notably of RR Lyrae stars, had yielded a good value for the distance to the center of the Galaxy, a value rather nicely confirmed by studies of galactic rotation. The heavy overlying obscuration made it impossible to photograph the center in normal blue or visual light. Some guesses could be made about the population characteristics of the nuclear region, which seemed to be mostly of Population II stars, but no solid information on the subject was available. Radio-astronomical evidence had already contributed much to our knowledge of the nucleus; a strong radio-continuum source had been located at precisely the position where we would expect to find the center from optical evidence. First attempts had been made to penetrate in the infrared through the dense obscuring matter that lies between the Sun and the nucleus of the Galaxy, but nothing very definite could be said.

All this has changed in the past 15 years. The nuclear region can now be studied in a variety of ways, especially by radio-astronomical and by infrared techniques, and there is a hint that x-ray observations may contribute further in the near future. Gravity waves emanating from the nuclear region have probably been detected and, if these observations are confirmed, they would seem to suggest that the nuclear region is losing mass at a fairly rapid rate. Observationally, the nuclear region of the Galaxy can now be studied in many different ways.

Interest in the study of nuclei of other galaxies has spurred on research about the nucleus of our own Galaxy. Following the early leads of the Soviet astronomer Ambartsumian,

there has come a realization that the nuclei of galaxies are the seats of major explosive events. Some of these are on most spectacular scales and involve the production of unbelievably large amounts of energy. Although the nucleus of our Galaxy seems to behave in a relatively moderate manner, it does show side effects of past and possibly of present explosions, the study of which can obviously contribute to the understanding of such phenomena in more distant galaxies. The nucleus of our Galaxy deserves a chapter of its own.

Optical Features of the Nuclear Region

The over-all picture of the surface distribution of faint stars along the band of the Milky Way directly supports the hypothesis that the star clouds in Sagittarius mark the direction toward the center of the Galaxy. The Palomar Schmidt photographs, which reach to very faint limits of apparent brightness, show that nowhere along the band of the Milky Way do the stars appear in greater numbers than in the Sagittarius section. In many ways it does seem surprising that it took until the time of Shapley's work before astronomers began to realize that the Sagittarius star clouds mark the direction to the center of our Galaxy. Observers in the Southern Hemisphere, who see the Sagittarius clouds of the Milky Way in all their glory, can hardly fail to be impressed by them. Nowhere does the Milky Way glow more brilliantly than in Sagittarius and Scorpius and nowhere else does it show such great apparent width. Telescopic views serve only to strengthen our conclusion that the Sagittarius clouds mark the direction toward the center. The evidence was there all the time, but no one took an unprejudiced look at the sky until Shapley pointed out some of the simple facts.

Unfortunately, relatively little is yet known about the detailed stellar composition of the nuclear region. The optical evidence, which does not reach to the nucleus itself (because of galactic obscuration), is very incomplete, but it suggests principally the presence of an agglomeration of older stars in the nuclear region. Radio-astronomical evidence shows that there is a thin and highly turbulent gaseous disk and a group of radio sources centered upon the nucleus itself. Strong infrared radiation is emitted by some objects in the central region.

The most direct evidence for the nuclear concentration of older stars related to Population II has come from studies of the distribution of RR Lyrae stars outside globular clusters. Walter Baade photographed a region about 4° from the direction of the galactic center that is remarkably free from local obscuration. He showed that for this particular field the total overlying photographic absorption over a distance of 10,000 parsecs is only a little more than 2 magnitudes. Baade's choice was an excellent one. A recent study by Van den Bergh shows only 1.5 magnitudes of foreground absorption overlying Baade's field. In this field, which marks the direction in which we can probably see closest to the galactic center, Baade found large numbers of RR Lyrae variables among the faint stars and, to his delight, he found a definite maximum in the numbers of these variable stars at apparent magnitude $m = 17.5$. This maximum is well above the plate limit, which is near $m = 20$. Correcting for the best available value of the total photographic absorption, and assuming a mean absolute magnitude $M_b = +0.5$, Arp finds that the maximum frequency of RR Lyrae stars occurs at a distance of 9,500 parsecs from the Sun. We note that this is very close to the distance of 10,000 parsecs between the Sun and the center of the

system of globular clusters. The value of the distance is slightly larger than the original one found by Baade and by S. Gaposchkin, who assisted Baade in the measurement and analysis of these variable stars. The search for and study of RR Lyrae variables should obviously be pressed for other relatively unobscured regions in the Sagittarius-Scorpius section. Such work is now under way under the direction of Plaut at the Kapteyn Astronomical Laboratory in Holland. The first results of these studies indicate the presence of RR Lyrae concentrations at distances less than the value generally assumed for the distance to the galactic center—8,500 rather than 10,000 parsecs.

Many varieties of stars of Population II show a marked concentration toward the center of the Galaxy. Long-period and other irregular variable stars, and especially also the novae, occur in greater abundance in the Sagittarius-Scorpius section of the Milky Way than anywhere else. Near-infrared surveys, conducted at the Warner and Swasey Observatory by Nassau, Van Albada, and Blanco, and at the Mexican Observatory in Tonanzintla by Haro and associates, have revealed that the late M-type giant stars (M5 and later) show little tendency toward clustering, but that they increase in numbers uniformly toward the center of the Galaxy. Comparable results have been obtained from Southern Hemisphere surveys by Henry and Elske Smith and by Westerlund.

Another group of objects that is of interest in this connection is the planetary nebulae. These nebulae possess gaseous shells of relatively small dimensions, which are often centered upon a hot, presumably old, star that may have been a nova in the past. The studies of Minkowski and of Perek on the distribution of these planetary nebulae have revealed that the fainter objects show a great preference for the Sagittarius-Scorpius section. Generally speaking, all stars and related objects that show clear Population II characteristics appear in unusual abundance in the section of the center of the Galaxy.

There is still considerable confusion about the exact stellar population that dominates the central region of our nearest neighbor, the Andromeda galaxy, Messier 31. The spectrum of the nucleus of Messier 31 was studied in 1957 by Morgan and Mayall, who found in it strong cyanogen bands, which are normally indicative of dwarf stars. Since then, Spinrad and associates of the University of California have done some very interesting analysis of the composite spectra of the nuclear regions of our Galaxy and of Messier 31. The great intensity of the sodium D lines in the spectrum of the center of Messier 31 would offhand seem to suggest that most of the red M stars are not giants, but rather dwarf stars. However, a more likely explanation is that the stars at the centers of our Galaxy and of Messier 31 are a special class of stars unusually rich in metallic lines. It is perhaps not surprising that we should find near the nucleus stars that are not composed principally of hydrogen and helium, but that apparently have been made from interstellar gas that had been subjected to atomic transformations early in the history of the Galaxy. We have noted already that explosive events on a large scale seem to be characteristic of nuclei of galaxies and it does not seem out of order to suggest that atom cooking on a large scale may have gone on long ago near the center of the Galaxy. All sorts of evidence shows that there is relatively less hydrogen gas right at the center than in regions at some distance from the center.

We should state clearly that the visible great star cloud in Sagittarius marks, in all probability, only the closer edge of a rather extended dense nuclear star cloud. As early as 1952, Bok and van Wijk demonstrated that in the direction of the galactic center there is very much local obscuring matter within 2,000 or 3,000 parsecs of the Sun. Modern estimates, which are based on infrared observations of the nucleus itself, yield approximate values of 25 to 30 magnitudes for the overlying obscuration in visible light between us and the galactic center. This means that only about one part in 1,000 billion of the light emitted by the center is observable in the visible part of the spectrum! Obviously we cannot hope to study the center by the standard techniques of astronomical photography.

The Radio Center of Our Galaxy

From the very day of the discovery that radio radiation is reaching us from distant parts of our Galaxy, it has been known that the central region of the Galaxy produces strong radio emission. Jansky's discovery paper of the middle 1930's gave conclusive evidence for radio radiation coming to us from the general direction of the galactic center. The center of the Galaxy figures prominently in the first radio-continuum maps, those published in 1944 by Grote Reber, which indicate a strong maximum in the direction of the Sagittarius clouds. In 1951 Piddington and Minnet showed that there is a strong discrete radio source in Sagittarius, which obviously marks the direction to the center. Following their suggestion, we refer to this source as Sagittarius A. Over the years, many radio maps of the central regions of our Galaxy have been made. One of the most famous of these, reproduced in Fig. 67, shows the radio equal-intensity contours for radio-continuum radiation with a wavelength of 3.75 centimeters observed by Downes, Maxwell, and Meeks. The

67. A radio map of the galactic-center region. This map, by Downes, Maxwell, and Meeks, shows the distribution of radio sources at a wavelength of 3.75 centimeters, obtained with the 120-foot Haystack Antenna of the Lincoln Laboratory, antenna beam width 4.2 minutes of arc. The contours of equal brightness represent antenna temperatures, which are directly proportional to the intensity measured at each point. The source G 0.7 0.0 is at galactic longitude 0°.7, galactic latitude 0°.0. (Courtesy of *Astrophysical Journal.*)

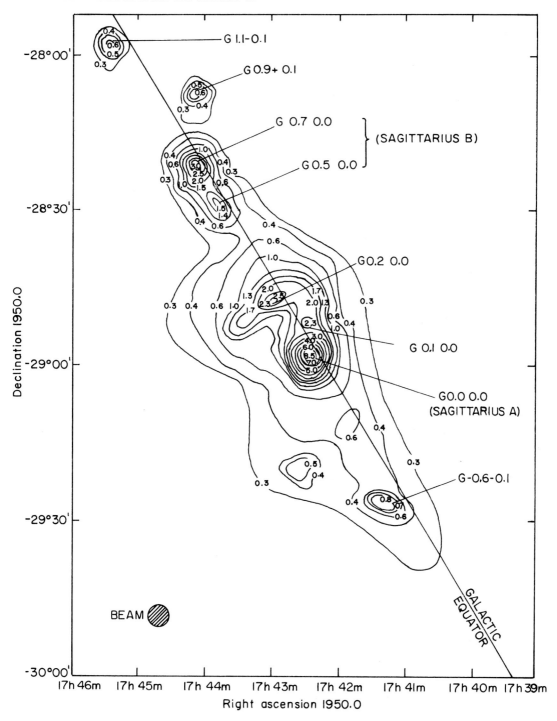

central region of the Galaxy is a conglomerate of a number of separate sources, each of which is labeled in Fig. 67 by its position in galactic longitude and latitude. The strongest of all is the source marked Sagittarius A, which is generally considered to indicate the precise position of the center of the Galaxy. The radio map contains seven named sources in addition to Sagittarius A.

There is a very interesting difference in the properties of Sagittarius A and of the seven radio sources that accompany it. When we study the relative brightness distributions of the seven other sources at different radio wavelengths, we find all seven of them to be relatively weak at longer radio wavelengths, whereas Sagittarius A has very strong radiation at these wavelengths. This alone suggests that there is quite a difference between the physics of the seven sources and the physics of Sagittarius A. An analysis of the brightness distribution with wavelength, that is, the spectra, of the seven sources shows that most of them radiate very much like the beautiful emission nebulae, to be discussed in Chapter 8, which produce their radiation by a simple process of recombination of free electrons with hydrogen nuclei. Why can we not photograph or see these radio sources? Simply because at optical wavelengths they are hidden from our view by the heavy overlying obscuration which, at visual wavelengths, amounts to between 25 and 30 magnitudes. The distribution of the continuum radiation from Sagittarius A is, however, very much like that found in distant extragalactic radio sources and in galactic supernova remnants, such as the Crab Nebula. This is the sort of radiation referred to as synchrotron radiation. It is produced only in the presence of strong magnetic fields. The radiators that produce the radiation are electrons spiraling at very high speeds around magnetic lines of force. The radio source, Sagittarius A, that marks the center of the Galaxy has all the earmarks of being an aftereffect from a mighty explosive event.

The radio fine structure of Sagittarius A has been investigated by Downes and Martin, also by Ekers and Lynden-Bell. It appears to be composed of two fairly large sources, Sagittarius A East, diameter 150 seconds of arc, and Sagittarius A West, diameter 45 seconds of arc, and a third compact source with a diameter of only 10 seconds of arc. The smallest detected unit measures less than 0.5 parsec in diameter!

Radio-continuum studies are only part of our story. For the past 20 years the central regions of the Galaxy have also been studied extensively at the radio wavelength of the 21-centimeter line emitted by neutral atomic hydrogen. This line originates from a transition in the neutral hydrogen atom from a level with very slightly more energy than the lowest level, and it is intrinsically a very sharp one; the neutral hydrogen atom at rest emits radiation at very nearly one wavelength only. If the hydrogen atom that emits the radiation is moving either toward or away from the Earth, the wavelength will be slightly shortened or lengthened according to the usual radial-velocity Doppler formula. If we observe 21-centimeter radio radiation of a given strength and at a wavelength differing by a certain amount from the normal one, then we know that this radiation originates from a cloud of neutral atomic hydrogen with a well-determined velocity of approach or recession relative to the observer. Modern radio telescopes are excellently equipped to study the radial-velocity shifts and intensities of the 21-centimeter radiation for any direction in the sky. The Dutch radio astronomers, notably Oort and Rougoor, have made most extensive studies of 21-centimeter radiation received from the section containing the galactic center. The most striking feature discovered by them at this wavelength is an *expanding spiral arm*, located at a distance of about 3,000 parsecs from the center of the Galaxy and expanding outward from it. The neutral atomic hydrogen that makes up the strong 21-centimeter radiators in the expanding arm, the section that is located between the Sun and Earth and the galactic center, comes toward us with a velocity of approach of about 50 kilometers per second. A second section of the arm can be seen beyond the center. It is moving away from us with a speed of 135 kilometers per second. It is not quite certain whether this whole feature represents a true spiral arm or a ring, but there is a very strong suggestion that it is expanding away from the center and that originally it was thrown out of the nucleus of our Galaxy. Assuming that the hydrogen in the feature was expelled by the nucleus, we can calculate that the original expulsion took place between 10 and 100 million years ago, which is a rather short time on the cosmic scale of time measurement, 0.04 to 0.4 of a cosmic year.

One of the strongest arguments to show that there is neutral atomic hydrogen between the Sun and Earth and the galactic center comes from the observation of 21-centimeter absorption lines viewed in the radio spectra of most of the sources shown in Fig. 67, including Sagittarius A. There are several other major 21-centimeter features associated with the section of the central region of the Galaxy. One of the most important of these is a rapidly rotating flattened nuclear disk of neutral atomic hydrogen with an outer diameter of about 800 parsecs, which whizzes

about the center at a rate of more than 200 kilometers per second. Furthermore, there is evidence of massive ejection of neutral atomic hydrogen in directions away from the central plane of the Galaxy, hydrogen clouds that can be observed at high galactic latitudes.

Some sections near the galactic center have proved to be happy hunting grounds for radio astronomers looking for new molecular lines. Sagittarius A shows strong absorption lines from the OH radical and from formaldehyde (H_2CO), both of which, from their radial-velocity shifts, give further proof of the turbulent nature of the gases observable in the central region of the Galaxy. There appear to be no molecules to speak of inside the Sagittarius A source; the absorption features are produced presumably in clouds located between the Earth and the galactic center. The molecules seem to prefer the conditions in the source Sagittarius B, which is located about 40 minutes of arc to the north and east of Sagittarius A. In Fig. 67 Sagittarius A is the strong source marked G 0.0 0.0, whereas Sagittarius B is marked G 0.7 0.0 and G 0.5 0.0.

We should note that x-ray observations are beginning to contribute to the study of the galactic nucleus. Giacconi, Gursky, Bradt, and others have found several x-ray sources in the section of the sky that contains the galactic center. An extended region of moderately strong x-radiation, 2 degrees in diameter, has been located centered upon Sagittarius A and the infrared sources that we shall describe later in this chapter.

Infrared Studies

In 1951, three Soviet astronomers, Kaliniak, Krassovsky, and Nikonov, were the first to penetrate in infrared light through the cosmic dust and to observe features associated with the galactic nucleus. Dufay and Berthier

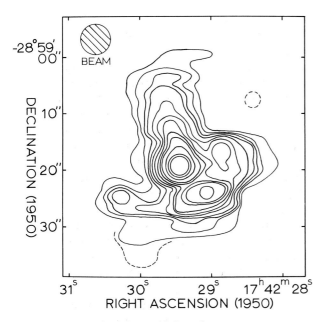

68. An infrared map of the galactic center at wavelength 10.5 microns. This map, prepared by Rieke and Low, covers only a very small region of the sky, as will be seen by comparison of the horizontal and vertical scales in this diagram with those in Fig. 67. It is restricted to the vicinity of Sagittarius A. The beam width for these observations was 5.5 seconds of arc. Contours of equal brightness are shown. There are at least four distinct sources. (Courtesy of *Astrophysical Journal*.)

in France discovered some new infrared features from their studies in 1959. A new era in infrared research began in 1968, when Becklin and Neugebauer found that infrared radiations from the galactic nucleus were observable at a wavelength of 2.2 microns (22,000 angstroms). Their strongest source has a diameter of 5 minutes of arc, which, at the distance of the galactic center, corresponds to about 15 parsecs. At its very center there is a small source with a diameter of only 15 seconds of arc, less than 1 parsec. This particular source seems to be close to the position of Sag-

ittarius A. These strong sources are viewed against an extended infrared background, and a number of additional discrete sources were indicated by this survey.

The central infrared source has been subjected to rather extensive investigation. Studies by Low, Kleinmann, Forbes, and Auman have shown that this source has terrific intensity at somewhat longer wavelengths, between 3 and 20 microns. Its integrated luminosity is the equivalent of 1 to 10 million Suns. A very striking result is that the intensity of the source at still longer far-infrared wavelengths, especially in the range between 30 and 300 microns, is far greater than one would expect from simple interpolation between the intensities observed in the near infrared and at the shortest radio wavelengths. One surprising result of the recent infrared work by Rieke and Low has been that there seems to be a small difference between the position of the radio source Sagittarius A and the strongest source in the infrared. Sagittarius A is located about 1.5 seconds of arc to the east of the infrared source. Figure 68 shows the detailed map for the direction of the galactic center by Rieke and Low. Their beamwidth is about 5 seconds of arc and at 10.5-microns wavelength their positions are accurate with a maximum uncertainty of 2 seconds of arc!

The identifying and measuring of infrared sources observable in the wavelength region between the shortest radio waves and the normal near infrared is not proving to be an easy task. Several additional infrared sources have apparently been found. An interesting current map has been prepared by Richard Capps (Fig. 69). It shows the principal radio and infrared sources in a single diagram. One feature included in this diagram is an extended source at 100-microns wavelength discovered

69. Infrared and radio sources near the galactic center. This map, prepared by Richard Capps, is a continuation of Fig. 67. The elongated contour is the result of observations at 100-microns wavelength by Hoffman, Frederick, and Emery. The positions of two secondary sources have been marked by squares. Broken circles give the positions of infrared sources observed at 2.2-microns wavelength by Becklin and Neugebauer. The sources detected by Auman and Low are marked by large crosses. (Courtesy of Kitt Peak National Observatory.)

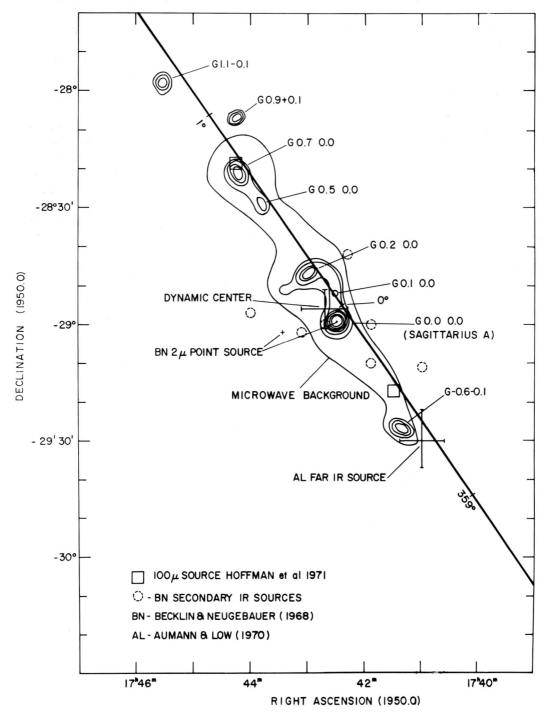

and studied by Hoffmann, Frederick, and Emery; it appears as a background against which the stronger sources are observed.

There are some striking similarities between the nucleus of our Galaxy and that of the Andromeda galaxy, Messier 31. In the infrared both galaxies have a sharply defined nuclear region. The intensity distribution at 2.2 microns within 40 parsecs of the center is very similar for the two galaxies. However, Messier 31 has no equivalent to the strong radio source Sagittarius A. The most detailed comparison between the two galaxies was made in 1970 by Rubin and Ford, who showed that dynamically and physically the two galaxies, including their nuclei, are quite similar to each other. For example, the sharp nucleus of Messier 31 has a diameter of about 2 seconds of arc, 8 parsecs, a value comparable to that of the infrared nucleus of our Galaxy. Spinrad's metal-rich giants may account for the near-infrared radiation of the source at 2.2-micron wavelength found by Becklin and Neugebauer.

There has been much speculation regarding the origin of the strong infrared radiation reaching us from the nucleus of the Galaxy. The similarities between the near-infrared properties (at 2.2 microns) of the center of our Galaxy and of Messier 31, taking into account the spectral properties found by Morgan and Mayall and by Spinrad, make it seem likely that large numbers of metal-rich giant stars are an important and massive component. As of now the simplest and most rea-sonable explanation for the strong infrared radiation coming directly from the nucleus of our Galaxy is that advanced by Low. He suggests that we are viewing here a complex of emission nebulae and hot stars embedded in dense clouds of cosmic dust. The infrared radiation we receive would then be the infrared radiation emitted by the dense dust clouds; we never see the hot stuff deep inside.

The study of the central regions of our Galaxy has only just begun. Whereas at first the region of the galactic center seemed to be mostly out of reach for the observer, it has now become accessible to observation and study at many wavelengths. For many years, it seemed that by optical means we would be limited wholly to studies of regions peripheral to the center itself. However, with the new infrared techniques, we have succeeded in penetrating through the cosmic dust and we have found what appears to be the real kernel of the central region. Radio data provide a wealth of material. We now have reliable and detailed maps of the sources of continuum radiation; 21-centimeter and molecular research are helping to sort out what is at the center and what lies in front and beyond. High-energy gamma radiation has now been detected coming from the band of the Milky Way and a broad maximum is shown for the direction of the galactic center. Gravitational waves are only just beginning to tell their story. The center obviously bears watching!

8
The Interstellar Gas

Emission Nebulae

Many of our readers have probably viewed the Great Nebula in Orion through a telescope. The soft greenish hue of the nebulous mass, gradually dimming toward the edge of the field, its erratic though immobile shadings, smooth and mellow in spots, hard and sharp elsewhere, together with the diamond-like scintillations of the four closely packed stars of the Trapezium, present a picture of unsurpassed beauty (Fig. 70). No telescope has ever been able to resolve this glowing mass into stars, and spectroscopic evidence shows that it is truly a nebulous cloud of gas, shining in the transmitted glory of its central stars. What causes such nebulae to shine?

The Orion Nebula has a bright-line spectrum in which the lines of hydrogen, ionized oxygen, and helium predominate. Nebulae like the Orion Nebula are not self-luminous. Nearly 50 years ago Hubble showed that a

very hot star is located in the immediate vicinity of every diffuse nebula that shows a spectrum similar to that of the Orion Nebula.

The physical theory that explains why and how such nebulae shine is basically quite simple. The densities and pressures in the nebulae are so low that according to earthly standards we would consider such regions to be perfect vacua. In our physical laboratories an atom is never left alone for any length of time; it is constantly bumping into one of its companions or into the walls of its container. If we wish to observe the atomic processes in their majestic simplicity, we turn to the diffuse nebulae, or to clouds of interstellar gas, which are apparently places where atoms are left alone long enough to perform without undue disturbances.

The atoms in the nebular gas are being bombarded in a leisurely fashion by the radiation from surrounding stars. The only light quanta that can produce major atomic excitation

70. The Orion Nebula. (Photograph made with
the Crossley reflector of Lick Observatory.)

are those of high frequency sent out in abundance by the blue-white stars of spectral types O and B. Most quanta of lower frequencies will simply filter through the gas without bothering it, or being bothered, to any appreciable extent. If a quantum of very high frequency, hence of short wavelength, strikes a neutral atom, it may transfer enough energy to the atom to cause the expulsion of an electron. The atom is then no longer electrically neutral, but positively charged. It became an *ion* by the process known as *photoionization*. The electron is free to start off by itself on a journey through interstellar space.

What can happen to the free electron? With its negative charge, it is ready to combine with any positively charged ion that is available, but it soon discovers that there are very few such ions. Our positively charged ion is equally hampered in its search for a free electron that would return it to the neutral state. An atom once ionized may travel through the nebula for days or months before it encounters a free electron to neutralize its charge. In the interstellar laboratories physical processes operate in unhurried and leisurely ways. An atom inside a star, or in one of the physical laboratories on the earth, is constantly being bumped and jostled. The atoms in the nebulae, however, live alone and like it.

Occasionally one of the free electrons will be captured by a positively charged ion. Let us suppose that a capture is made by a hydrogen ion, that is, a *proton*. According to modern atomic theory, the neutral hydrogen atom has only a limited number of permitted orbits in which the electron can move about the proton (Fig. 71). Each such orbit has a definite energy associated with it. When an electron is captured by the proton, it can land in any one of the permitted orbits of the now

neutral hydrogen atom. If the capture takes place in the tightest orbit—that of lowest energy—the whole show will be over at once; a single ultraviolet quantum will then be emitted as a result of the capture. Frequently, however, the free electron will be captured in one of the orbits of higher energy. The hydrogen atom cannot remain for more than a small fraction of a second in this excited state and the capture is followed almost immediately by a series of stepwise transitions of the captured electron from one orbit to another of lower energy. The electron cascades inward to end up in the orbit of lowest energy, where it will remain until the next ultraviolet quantum comes along to begin another sequence of disruption and ultimate recombination. During the cascading process a light quantum will be released as the result of each transition. The chances are that somewhere on the way a quantum corresponding to one of the Balmer transitions will be produced which involves a transition from a higher level to the basic level with quantum number $n = 2$. The Balmer lines in diffuse nebulae are produced during such internal adjustments in the neutral atom, which follow immediately after the capture of a free electron by a proton.

The beautiful and conspicuous emission nebulae such as the Great Nebula in Orion, or the famous nebula near the star Eta Carinae, are not isolated phenomena; they represent marked concentrations of interstellar gas, which occupy much of the space between the stars in and near the central plane of our Milky Way system, especially in the spiral arms. There are various ways in which this gaseous substratum reveals itself to the observer. Modern photographic and spectrographic techniques have enabled us to record the faint outer extensions of the larger nebulae

71. Emission of Balmer lines by diffuse nebulae. The circles represent the relative energy levels of the electron in the hydrogen atom, with 1 indicating the lowest or Lyman level, 2 the Balmer level, 3 the Paschen level, and so on. The Balmer series and the Lyman series are shown by arrows representing transitions from higher levels to the Balmer and Lyman levels, respectively. The lines A, B, and C show ways in which an electron may be captured by a proton to produce a neutral atom in the Balmer level. In A the electron is captured immediately in the Balmer level and as a result a quantum in the Balmer continuum is emitted. In B and C the capture takes place in level 5 with the emission of an infrared quantum in the continuum for level 5. In B this capture is followed by a direct transition to the Balmer level and emission of the H-gamma line. In C the first transition is from level 5 to level 3 and then follows a transition to the Balmer level with the emission of the H-alpha line. The electron in the hydrogen atom would remain in the Balmer level for 1 one-hundred millionth of a second and then settle down in the lowest or Lyman level.

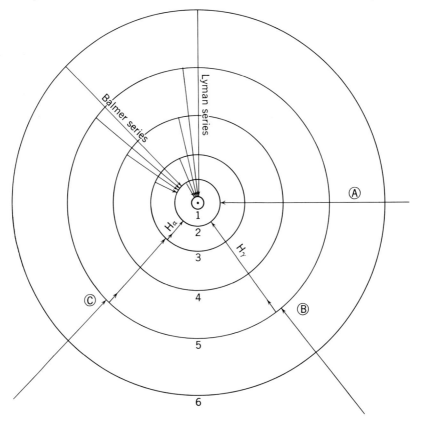

and to locate sometimes very extensive and very faint emission nebulae that had not been previously observed. The interstellar gas is also detected through the presence of certain sharp absorption lines observed in the spectra of distant stars. And, finally, fresh evidence is now coming from radio astronomy, and most recently from the Orbiting Astronomical Observatories. Added interest in studies of emission nebulae arises through the discovery by Baade that the spiral arms in remote galaxies outside our own Milky Way system are most readily traced through alignments of patches of emission nebulosity. If we can locate the emission nebulae of our own Milky Way system and determine their true distances from the Earth, then we should be able to trace the spiral arms of our Galaxy.

Abundances of the Chemical Elements in Nebulae

One might almost gain the impression from our introductory section that the only chemical element present in interstellar space and in emission nebulae is atomic hydrogen. This is not so. Helium atoms are present in considerable abundance, and the appearance of nebular spectral lines attributable to various stages of ionization of nitrogen, oxygen, neon, and some of the heavier elements shows that other elements are present as well. The first estimates of relative abundances of the chemical elements in interstellar space were made by Dunham, and improved values were published subsequently by Strömgren and by Seaton. There have been many recent determinations of relative abundances of the chemical elements, mostly by Aller and his associates, D. J. Faulkner and Helene Dickel, by Mathis, and by H. M. Johnson. Although there are minor differences from one nebula to another, one can say that, on the average,

for every 10,000 hydrogen atoms, there are present 1,200 helium atoms, 1 or 2 nitrogen atoms, 3 or 4 oxygen atoms, 1 neon atom, and 1 sulfur atom, with smaller numbers of some heavier atoms, such as iron and chlorine. The casual chemical analyst in interstellar space would conclude that the interstellar gas is mostly a mixture of hydrogen and helium atoms, with a trace of impurities present!

Rather similar relative abundances of the chemical elements are found in the emission nebulae of the Large Magellanic Cloud and in the nearby spiral galaxies Messier 31 and Messier 33. It is of interest to compare the relative abundances of the elements inside stars and in emission nebulae and the interstellar gas. Early in 1973, Spitzer, Morton, and collaborators reported that heavy elements such as magnesium, phosphorus, chlorine, and manganese are less abundant in interstellar space than in the Sun by factors ranging between 4 and 10. Other heavy elements seem to be similarly depleted relative to the Sun. These results were obtained on the basis of observations made from Orbiting Astronomical Observatory No. 3, the famous Copernican Satellite. The observations suggest that heavy elements may have a tendency to stick to the cosmic grains more readily than light ones.

There is proportionately far less hydrogen and helium on Earth than there is in the Sun or in interstellar space. The reason for this can be readily understood. At the time when our Earth was formed, the light hydrogen and most of the helium atoms moved fast enough to escape from the gravitational pull of the proto-Earth. The heavier atoms did not escape. The only hydrogen that stayed with us was that already combined into molecules that were heavy enough to be retained by the crust and the atmosphere. A trace of helium remains in the upper atmosphere of the Earth.

72. The Orion Nebula in infrared light. The photograph shows clearly a marked concentration of very red stars in the region of the gaseous nebula. Comparison with Fig. 70 shows many interesting features. The sharp lines emanating from the star are diffraction patterns produced by the various supports in the tube of the telescope. The white ring around the black circle results from reflection of the light of the star against the glass back of the photographic plate. (An enlargement from a hypersensitized infrared plate by Haro made with the Tonanzintla Schmidt telescope.)

73. Objective-prism spectra for a region near the Orion Nebula. The region is that shown in the lower right-hand corner of Figs. 70 and 72 (tilted). A red-sensitive emulsion was used for this photograph by Haro at the Tonanzintla Observatory. The marks (by Haro) indicate stars with the Balmer H-alpha line in emission.

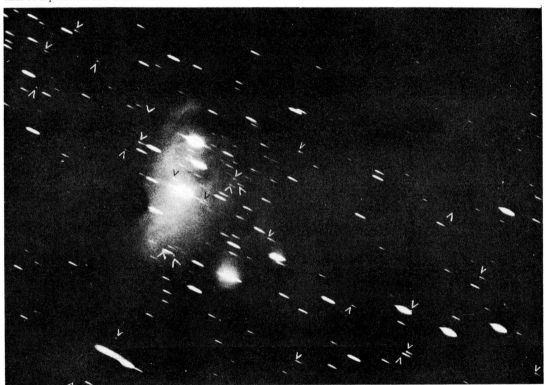

Optical Studies of Emission Nebulae

In 1937, Struve and Elvey constructed at the McDonald Observatory in Texas a mountainside spectrograph, capable of recording spectral lines of very faint and extended emission nebulosities which, before that time, had not been known to exist. More recently it has become possible to photograph these same nebulae, and other hitherto undetected nebulae, with the aid of fast red-sensitive photographic emulsions and special color filters. Direct filter photography is basically a very simple process. Earlier attempts at direct photography of very faint emission nebulae had failed because the luminous background of the sky blackened the photographic plates before the faint nebulae had registered themselves sufficiently for detection. Astronomers reasoned that, if they could only suppress the sky fog without cutting out much of the nebular light, they should be able to take very long exposures with fast cameras and so record the faint nebulae. It should be remembered that the nebulae shine in the light of specific spectral lines, but that the sky fog on our photographs is produced mostly by the aggregate light of all wavelengths from thousands of faint stars and by certain specific radiations present in the background light of the night sky. By a proper selection of color filters with narrow transmission bands, we are able to cut out all but a small fraction of the continuous light from the background stars and also to suppress most of the night-sky radiation, while permitting the light from one of the nebular emission lines to pass practically undiminished (Fig. 74). In practice this selective photography is achieved most readily through the use of the hydrogen emission line, H-alpha, with a wavelength of 6,563 angstrom units, in the red part of the spectrum.

The most effective filter is a Corning red filter combined with one of the many available narrow-band interference filters. One type of filter that has thus far been successfully employed in work of this sort transmits close to 90 percent of the H-alpha light of the nebula, but almost no radiation is transmitted outside a band 50 angstroms wide centered on H-alpha. With such a filter arrangement, and with the fastest red-sensitive plates produced by Eastman Kodak, it has been found practicable to make exposures of up to 4 hours with a camera operating at a focal ratio of 1.5.

The search for faint, extended nebulosities has proved so rewarding that in recent years many observatories have participated in it. In the Soviet Union, Shajn and Miss Hase at the Simeis Observatory in the Crimea and Fessenkov at Alma Ata were the pioneers of photography showing filamentary structure overlying large sections of the northern Milky Way. In France, Courtès and Dufay worked in the same field, and in the United States some of the finest photographs have been made at the Yerkes and the McDonald Observatories, first in 1952 by W. W. Morgan, Sharpless, and Osterbrock and later by Morgan, Strömgren, and H. M. Johnson. The southern Milky Way has also received its share of attention; the first surveys for the half of the Milky Way that is richest in emission nebulosity were made by Gum in Australia and by Bok, Bester, and Wade and also by Code and Houck from the Boyden Station in South Africa. The finest available atlas of the southern Milky Way is the *Mount Stromlo H-Alpha Atlas*, a joint effort by Rodgers, Campbell, Whiteoak, Bailey, and Hunt. It covers a band 30° wide of the southern Milky Way from Sirius past the Scutum Cloud (Fig. 3). There exists, as yet, no comparable atlas for the northern Milky Way. All

74. A matched pair of photographs of the region of the North America Nebula. The left-hand photograph was made on a red-sensitive emulsion and represents the nebula almost entirely in hydrogen H-alpha light. The right-hand photograph was made on an infrared-sensitive emulsion and with a filter to exclude H-alpha light. The total absence of bright nebulosity in the right-hand photograph is to be noted. We draw attention to the fact that the shape of the North American continent is produced by dark nebulosity, especially in the "Gulf of Mexico" and in the "North Atlantic." The famous Pelican Nebula is shown in the left-hand photograph to the right of the North America Nebula. (Photographs by Dufay at the Haute Provence Observatory.)

75. A bright fan-shaped nebula in Carina. The southern gaseous nebula NGC 3581, photographed in red light with the ADH telescope of the Boyden Observatory.

76(*a*). The Milky Way in Scorpius in H-alpha
light. The key chart shown in Fig. 76(*b*) indicates
the positions of the principal gaseous nebulae in
this section of the Milky Way. These nebulae are
part of an inner spiral arm of our Milky Way system. (Perkin-Zeiss camera, Boyden Observatory.)

76(b). Key chart for Fig. 76(a). The full lines indicate the principal emission nebulosities shown in Fig. 76(a); the dotted line gives the outer boundary of detectable nebulosity. The numbers refer to the positions of the features in galactic coordinates under the former system of galactic coordinates; to refer to the system now in use, 33° should be added to the given galactic longitudes. The first three figures are the longitudes, the last two the latitudes; underlining indicates negative latitudes. A section of the Sagittarius spiral arm can be traced diagonally from upper left to lower right in the diagram.

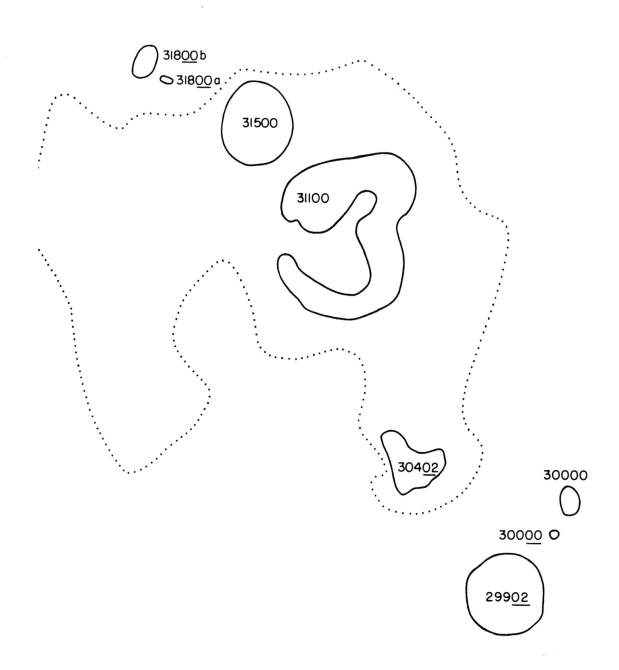

NORTH

77. The Milky Way in Scorpius; H-alpha light excluded. The photograph covers precisely the same area of the Milky Way shown in Fig. 76(a), but here a narrow-band Baird filter was used, selected to exclude practically all H-alpha radiation. The emission nebulae are either absent or very weak. (Perkin-Zeiss camera, Boyden Observatory.)

this work with relatively small wide-angle cameras has resulted in a fairly complete mapping of the sections in which emission nebulosity, including very weak emission, occurs over rather large areas of the sky.

The linear scales of the photographs made with the small search cameras are, however, too restrictive to reveal the full detail of each nebulous structure, so that some of the finer features may remain undetected. Hence large telescopes with fairly small focal ratios (and therefore of great light-gathering power) are used to bring out the details. In this connection the 48-inch Schmidt telescope of the Mount Palomar Observatory and the 32-inch Baker-Schmidt telescope of the Boyden Observatory, known as the Armagh-Dunsink-Harvard telescope, have proved very helpful for the charting of the northern and southern Milky Way. The 24-inch Curtis-Schmidt telescope of the University of Michigan, now at Cerro Tololo Inter-American Observatory in Chile, and the 20-inch Swedish Uppsala-Schmidt telescope at Mount Stromlo Observatory in Australia, have done effective work on wide-angle photography for the Southern Hemisphere. Two additional Schmidt telescopes of large aperture and one wide-field reflector are being added to the southern arsenal: the 40-inch Schmidt for the European Southern Observatory in La Silla in Chile, the 48-inch British Schmidt for Siding Spring Observatory in Australia, and the 100-inch reflector for the Hale Observatory at Las Campanas in Chile.

There exists as yet no complete catalog of emission nebulae found from surveys made with the southern Schmidt telescopes, but considerable work has been done on the basis of the Palomar Schmidt survey. The most comprehensive catalog was published by B. T. Lynds, who was then at Steward Observatory.

Her list contains close to 1,200 entries and covers all except the southernmost quadrant of the band of the Milky Way (galactic longitude 250° to 340°). A Schmidt photograph covers an area of the sky about 100 times that covered by the full Moon, and yet the scale of the photograph is sufficient to show details of structure in the nebulous filaments.

To round out the telescope picture for photography of emission nebulae, we turn finally to the large reflectors, which show the more conspicuous nebulae in all their glory. The 200-inch Hale and the 100-inch Hooker telescopes of the Mount Wilson and Palomar Observatories and the 120-inch Lick reflector are the best available telescopes for photography of nebulae from the Northern Hemisphere. The 74-inch reflectors of the Radcliffe Observatory in South Africa and of Mount Stromlo Observatory in Australia are the major instruments of the Southern Hemisphere. Four additional large reflecting telescopes are now under construction. The 158-inch Mayall reflector of Kitt Peak National Observatory is ready for Northern Hemisphere research. Three major telescopes are under construction for the Southern Hemisphere; they are the 158-inch reflector for Cerro Tololo Inter-American Observatory in Chile, the 150-inch British-Australian reflector for Siding Spring Observatory in Australia, and the 140-inch reflector for the European Southern Observatory in Chile. We are entitled to have great expectations!

Thus far the optical searches for faint, extended emission regions have concentrated largely upon nebulosities shining in the red light of the H-alpha line of hydrogen. We noted at the beginning of the present chapter that the H-alpha line originates as the result of the capture of a free electron by a single proton. Since this is the dominant process,

78. The Norma Nebula. (From a photograph by Westerlund with the Uppsala-Schmidt telescope.)

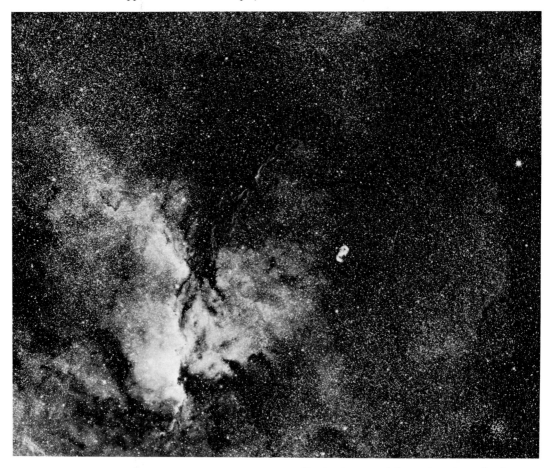

the observation of H-alpha is indicative of the presence of ionized hydrogen atoms (protons) caught in the act of recombination. In the professional jargon of the astrophysicist, the clouds of hydrogen that are largely ionized are called H II regions. The term was first introduced by Strömgren, who showed that H II regions should occur around O and B stars rich in ultraviolet radiation capable of ionizing the nearby interstellar hydrogen. Strömgren finds that, for an average density of interstellar hydrogen of the order of one hydrogen atom per cubic centimeter, a B0 star emits sufficient ultraviolet radiation to ionize all the hydrogen to a distance of 30 parsecs from the star. For the same hydrogen density, a very hot O star may produce complete ionization within a sphere of radius 200 parsecs, resulting in a hydrogen emission nebula of truly gigantic size. The supply of ultraviolet light and the resultant ionizing power decrease rapidly as we proceed to the cooler stars. An A0 star, with a still rather respectable surface temperature of the order of 11,000°K, will probably ionize the gas within a sphere with a radius of only 0.3 parsec, and the cooler stars will produce no appreciable hydrogen ionization at all. It therefore causes little surprise that near almost every H II region one finds an O or B star, or a cluster of O and B stars, that may be held responsible for exciting the nebular radiation.

We are really concerned with two varieties of regions of interstellar hydrogen, the H II regions and the H I regions. The latter contain principally neutral atomic hydrogen, whereas in the H II regions the majority of the hydrogen atoms are ionized. In the H II regions, the ionizing ultraviolet radiation from nearby stars is relatively plentiful, and the electrons are ejected at the time of ionization with sufficiently high speeds to produce temperatures in the interstellar gas of the order of 10,000°K. In the H I regions ionization does not play much of a role and there is no process by which interstellar atoms may maintain the same high speeds as for the H II regions. According to estimates by Spitzer and Savedoff, the speeds of the interstellar atoms in H I regions correspond to a temperature of 60°K, that is, 60° above the absolute zero, or, on the centigrade scale, more than 200° below the freezing point of water.

In recent years, increasing emphasis has been placed upon studies of radial velocities of emission nebulae and of spreads in radial velocities inside these nebulae. The standard approach in optical radial-velocity studies has long been to measure radial velocities from slit spectra obtained with large reflectors and refractors. The original Lick Observatory survey by Wright was carried out by these techniques. As spectrograph designs improved, it became possible to obtain radial-velocity data

for increasingly less conspicuous H II regions. One of the finest studies in this area done in the middle 1960's, is that of J. S. Miller of the University of Wisconsin, who used a fast slit spectrograph attached to the 36-inch Kitt Peak reflector. He obtained data for 36 H II regions, mostly of the northern Milky Way. Such radial-velocity data figure prominently in studies of the spiral structure of our Galaxy.

Optical measurements of faint H II regions have become possible by the application of the Fabry-Perot techniques of interference-fringe photography. These techniques have been developed for studies of H II regions principally by Courtès and his associates Cruvellier, Monnet, and the Georgelins. Most of the work was done in France at the Haute Provence Observatory, some of it at an observing station in South Africa, and more recently at the European Southern Observatory in Chile. More than 4,000 individual radial velocities have been measured, and a list of radial velocities is now available for 200 H II regions. The photographic techniques developed and applied by the French astronomers are now being used by M. G. Smith at Kitt Peak and Cerro Tololo observatories for photoelectric analysis. A pressure-scanning technique enables the observer to obtain directly at the telescope the intensity distribution in the profile of an emission line, which not only yields an average radial velocity of the nebula, but also provides data on the distribution of radial velocities within the part of the nebula being scrutinized. The average radial velocity represents very useful kinematical information for the study of spiral structure of our Galaxy, whereas from the spreads of radial velocity we may learn much about the physical conditions caused by the exciting star inside the nebula.

Interstellar Absorption Lines

The interstellar gas reveals itself not only through the characteristic bright-line spectra of the emission nebulae, but also through certain sharp and narrow absorption lines found in the spectra of many distant stars. The discovery of such absorption lines goes back to 1904, when the German astronomer Hartmann showed that the absorption K line of ionized calcium (wavelength 3,933 angstroms) in the spectrum of the star Delta Orionis behaved in a very peculiar fashion. Delta Orionis is a blue star with B0 spectrum and was recognized to be a spectroscopic binary. Hartmann found, however, that the wavelength of the K line did not vary at all in the course of the binary period. The hydrogen and helium lines in the spectra of Delta Orionis were broad and fuzzy, but its K line was sharp and distinct. Hartmann referred to the line as the "stationary" calcium line, since it did not share in the radial-velocity shifts exhibited by the spectral lines in the binary star.

Stationary calcium lines have since been discovered in the spectra of many other early-type stars. In 1919 Miss Heger (Mrs. C. D. Shane) and Wright at the Lick Observatory found that the spectra of some early-type stars have strong stationary sodium lines in addition to their stationary calcium lines. In all cases these lines were found to be sharp and distinct.

The interstellar origin of these absorption lines was first suggested by V. M. Slipher in 1909, but his suggestion unfortunately did not receive the attention it deserved and his views were not generally accepted until more than 15 years later. It was thought at first that the stationary lines originated in the immediate vicinity of the stars in whose spectra they are found. The researches of J. S. Plaskett and of Struve proved that this explanation was incorrect. Theoretical investigations, especially those of Eddington and Rosseland, showed also that stationary lines of calcium would naturally be observed if any free gas were present in interstellar space. The total evidence left little doubt as to the interstellar origin of the stationary lines and, since 1930, they have generally been referred to as interstellar lines.

Adams and Dunham at the Mount Wilson Observatory found interstellar absorption lines attributable to neutral calcium, neutral potassium, neutral iron, and ionized titanium in addition to the lines of ionized calcium and neutral sodium to which we have already referred (Figs. 79 and 80); all told, 15 sharp interstellar absorption lines were definitely identified. Dunham also found some sharp interstellar lines in the blue-violet part of the spectrum which for some time remained unidentified. McKellar of the Dominion Astrophysical Observatory in Canada proved that these lines arose from transitions between energy levels in simple molecular compounds of carbon, nitrogen, and hydrogen. In a recent tabulation, G. Münch lists 13 lines of molecular origin as definitely identified. Finally, Merrill of the Mount Wilson Observatory has given evidence of the presence of 6 broad and as yet unidentified interstellar absorption bands, the strongest at 4,430 angstroms with an approximate width of more than 40 angstroms. It is not yet known how these broad bands are produced. The suggestion has been made that they may come from more complex molecules or from a transition stage in the formation of solid particles. But for the time being all this is frankly guesswork.

The far ultraviolet spectra of half a dozen stars observed by the Copernican Satellite

79. Multiple interstellar lines. High-dispersion spectra made by Adams at the Mount Wilson Observatory show many instances of multiple interstellar lines. This multiplicity indicates the presence of distinct interstellar clouds, each moving at its own peculiar rate.

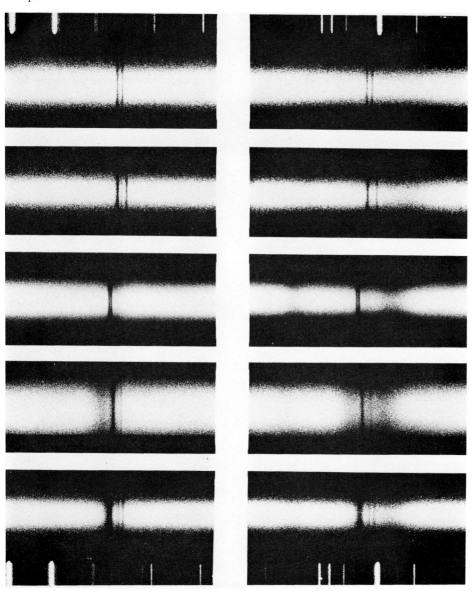

80. Interstellar lines of ionized calcium and of ionized CH. The marked sharp lines are examples of interstellar absorption lines—several of them multiple lines—discovered by Adams. (Mount Wilson photograph.)

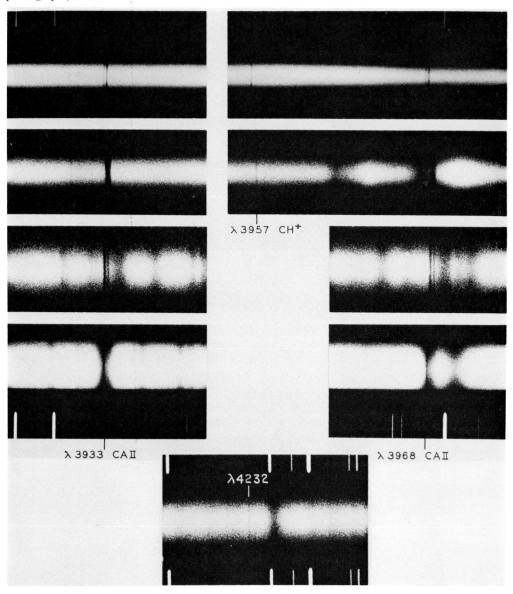

have yielded a harvest of new interstellar absorption lines. The lines produced by various ionization stages of carbon, nitrogen, oxygen, magnesium, silicon, phosphorus, sulfur (four ionization stages observed!), chlorine, argon, manganese, and iron have been identified by the Princeton group headed by Spitzer and Morton. Many absorption lines attributable to the hydrogen molecule (H_2) are present. This molecule was first detected by Carruthers in 1970 on the basis of rocket ultraviolet observations. The Copernican Satellite spectra show many H_2 absorption lines. In addition heavy hydrogen or deuterium (D) is found from telltale lines apparently produced by the HD molecule. The data indicate a surprisingly great abundance of D compared to H, the indicated ratio being that for every D nucleus there are 200 H nuclei, which represents a very much larger deuterium abundance relative to hydrogen than is found on Earth or in the Sun's atmosphere.

How do the observed interstellar absorption lines originate? Suppose we consider the interstellar *K* line with a wavelength of 3,933 angstroms. This absorption line is produced when an ionized calcium atom absorbs a light quantum of that wavelength. Since atomic phenomena take place in a rather leisurely fashion in interstellar space, we may assume initially that most ionized calcium atoms will have settled down to the lowest energy level. Then there may by chance come along a stellar quantum of wavelength 3,933 angstroms, just capable of temporarily exciting the calcium ion into a higher level of energy. Atomic physics teaches us that the calcium ion can stay in that blown-up state for only something like 1 ten-millionth of a second and that then it returns to its original lowest energy level. The energy that was absorbed is thereby released, but the new quantum of

wavelength 3,933 angstroms will go off—scatter—in a direction different from that in which the previous one arrived. If we are looking at the spectrum of a certain star and there is ionized calcium between that star and us, the interstellar calcium ions will absorb and scatter many of the quanta of wavelength 3,933 angstroms that would otherwise have reached our spectrograph; hence a dark line appears in the spectrum of the star.

The traditional technique for the study of interstellar absorption lines has been through the study of spectra of very high dispersion obtained with the aid of coudé spectrographs attached to large reflectors. The most active research worker in the field has been G. Münch at the Hale Observatories. Together with Vaughan, he has developed a high-resolution photoelectric Fabry-Perot system, which shows many components in each interstellar absorption line. Another approach to the same problem has been that of C. R. Lynds and Livingston, who used the McMath Solar Telescope at Kitt Peak for high-resolution measurements of the interstellar absorption lines for some bright stars within reach of this equipment.

The interstellar gas shows considerable cloud structure. Until 1936, astrophysicists supposed that the interstellar gas was distributed smoothly through a thin layer near the central plane of the Milky Way. In that year Beals, then of the Dominion Astrophysical Observatory in Canada, found evidence of multiplicity in some of the lines, indicating that more than one cloud of gas was contributing to the formation of the interstellar absorption lines in certain stars. Subsequent studies at high dispersion of the *K* line of ionized calcium were carried out by Adams at the Mount Wilson Observatory. He showed that more than 30 percent of the stars ex-

81. Interstellar sodium lines. The spectrum of star No. 12953 in the Henry Draper *Catalogue* photographed by Münch with the coudé spectrograph of the 200-inch Hale telescope. The black central band is the negative impression of the stellar spectrum and the white lines in the dark band are two components of the D_1 line of neutral sodium and two components of the D_2 line also due to neutral sodium. There are two components to each line because between the star and our Sun there are two clouds producing neutral sodium absorption, one of which, with a radial velocity of approach of 55 kilometers per second, gives the weaker component (to the left) of each pair, the other, with a radial velocity of approach of 7 kilometers per second, yielding the stronger component (to the right) of each pair. The lines at the top and bottom are from the comparison spectrum imprinted during the exposure. They are emission lines and hence appear black in the negative print. The star is of spectral class A1 and luminosity class Ia. It is at an estimated distance of 2,000 parsecs, and it so happens that, in the limited stretch of spectrum shown, there are no stellar absorption lines present.

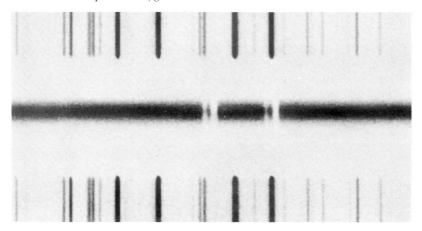

amined by him had double or triple interstellar K lines, and a K line with four components was found for four stars. It may be assumed that every observed component of the K line is produced by a separate and distinct interstellar cloud. The separation between the components is caused by the difference in the Doppler shifts produced by different radial velocities of the separate clouds relative to the Sun.

In recent years, G. Münch has found many stars that show multiple interstellar lines (Fig. 81); some of his spectra show as many as seven components, every one of them indicative of a separate interstellar cloud with a different radial velocity relative to the Sun. Münch's observations present most convincing evidence for the cloud structure of the interstellar gaseous medium. We note here that these multiple interstellar lines found by Münch occur even in the spectra of some stars at considerable distances from the galactic plane, thus suggesting that at least some wisps of interstellar gas have sufficient speeds at right angles to the plane of the Milky Way to reach considerable heights above or below the plane. We shall see in Chapter 11 that the great majority of the interstellar clouds discovered by Münch are near the central plane and fit nicely into the accepted spiral pattern of our Galaxy.

The interstellar gas partakes in the general

rotation of the Galaxy. This fact was established by the researches of Plaskett and Pearce at the Dominion Astrophysical Observatory. Shortly after Oort had suggested that the radial velocities of distant stars should vary in a double sine wave according to galactic longitude, Plaskett and Pearce undertook to measure the radial velocities of several hundred faint, and hence distant, early-type stars and they made a successful check of the theory of galactic rotation. The interstellar K line was measurable on many of the spectrograms of Plaskett and Pearce. When the radial velocities determined from the measurements of the interstellar K lines were plotted against the galactic longitude of the stars, the results showed very clearly the familiar effects of galactic rotation. The striking difference was that the range of the double sine wave with galactic longitude for the stars was approximately twice that shown by the radial velocities from the K line. The conclusion that Plaskett and Pearce drew from their curves was that the interstellar K line in the spectrum of a given star yields on the average a radial velocity corresponding to the halfway point between the star and the observer.

The early measurements based on interstellar lines seemed to suggest that interstellar calcium atoms are distributed rather uniformly near the galactic plane. Because of the discovery of multiple interstellar absorption lines in the spectra of many stars, this simple picture has now been abandoned. We visualize the interstellar gaseous medium as consisting primarily of a relatively flat layer of individual gas clouds, with a thin all-pervasive galactic substratum possibly connecting them. Each cloud moves around the center of the Galaxy in an approximately circular orbit. On the average the actual velocity of a cloud differs from the circular velocity of ga-

lactic rotation by ± 8 kilometers per second. With very few exceptions—which we shall note later on—the clouds that produce the observed interstellar absorption lines are located in the spiral arm that contains our Sun.

The average properties of the individual interstellar clouds can be found from a study of the components of the interstellar absorption lines shown by stars at various distances. On the average, a line of sight close to the galactic plane cuts through seven or eight of these clouds in 1 kiloparsec. Each cloud has a diameter of between 10 and 15 parsecs and a probable mass equivalent to a few hundred solar masses.

Radio Studies of the Interstellar Gas

During the past decade, radio techniques for the study of emission nebulae have become increasingly important. First of all, we learn from such studies much that is new about the physics of the nebulae, and second, we find some of the more distant emission nebulae very convenient anchor points for studies of remote spiral features in our Galaxy. These radio studies supplement in many ways the optical work on emission nebulae. One of the very useful properties of radio radiation is that it passes almost undiminished in intensity through clouds of cosmic dust. Many emission nebulae that are hidden from direct view by interstellar absorption can be detected and observed with ease by radio techniques. This is becoming increasingly important. Not only does it permit access to distant nebulae that otherwise would remain unknown, but it gives us an opportunity as well to detect and observe some relatively nearby and also many distant emission nebulae that are embedded in clouds of cosmic dust. It has become increasingly evident in recent years that protostars and other very

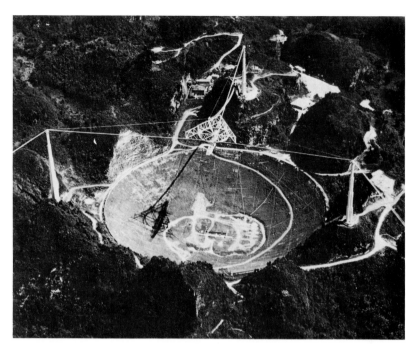

(a)

(b)

82. Four photographs of the 1,000-foot radio telescope at Arecibo, Puerto Rico. Figure 82(a) shows how advantage has been taken of the terrain to mount the 1,000-foot reflector firmly so that it can reflect radio beams from small areas of the sky to the receivers at the focus. The complex arrangements at the focus are shown in (b). Various focus probes are shown; each of these can be placed in the proper position at the precise focus for receiving and recording the radiation at the radio frequency in the range which that particular probe is capable of receiving. To improve the quality of radio reception at high radio frequencies (short wavelengths) the reflector itself was resurfaced with precision panels. Figures 82 (c) and (d) show the resurfacing work in progress during mid-1973. At the time these photographs were taken, about half the surface had been covered with the precision panels; the work was completed late in 1973. Note in (b) the size of the observer in the cage being transported to the focal area, and in (c) the size of the technician working on one of the panels. The focus probe can remain locked on one position for about an hour; the control is by computer. With its new reflector cover, the Arecibo Radio Telescope is one of the most powerful instruments for study of the interstellar gas and of pulsars and supernova remnants. (Courtesy of National Radio Astronomy and Ionosphere Center, Cornell University.)

(c)

(d)

young stars have associated with them considerable amounts of interstellar gas. This leads to the formation of emission nebulae, to which we have referred already as H II regions. However, such young star-gas configurations are often embedded in thick and dense clouds of small solid particles, which do not permit us to view the objects at normal optical wavelengths. Radio radiation is not bothered by this cosmic dust, and the young grouping, or at least the gas associated with it, can be studied nicely at radio wavelengths. We note here that some of these dust-embedded emission nebulae often also can be detected through observations in the far infrared.

The oldest technique for studying emission nebulae by radio methods makes use of the effects produced by free-free transitions. In H II regions there is of course much ionized hydrogen. The passage of a free electron close to a positively charged nucleus of the hydrogen atom, the proton, produces continuum radiation in the centimeter and decimeter range. The observed strength of the radiation depends on the number density of the protons and electrons and on the speeds with which the free electrons move about. These speeds are of course fixed by the temperature of the free-electron gas. The continuum radiation was first observed by radio astronomers in the early 1950's; by now the whole of the band of the Milky Way, north and south, has been mapped for continuum radiation, especially by Westerhout.

A second way in which ionized hydrogen can be studied by radio techniques is through the so-called alpha transitions, a method first suggested by Kardashev of Moscow. In regions with much ionized hydrogen, the capture of a free electron by a proton often will take place in one of the very high-energy levels of the neutral hydrogen atom. After capture, as we have mentioned, the electron will stay in this high level for only a minute fraction of a second, and it will then cascade down toward levels of lower energy, ultimately ending up in the level of lowest energy of the hydrogen atom, the Lyman level. The first transitions in the cascading process may well be from, say, level 158 to 157, or from 110 to 109. Happily, many such high-level transitions fall within the wavelength and frequency ranges of modern radio-astronomical receiving equipment. What we observe in each case is a quite sharp radio emission line, for which the rest wavelength is very accurately known. Any shift in frequency between the observed and rest frequencies may be attributed to radial velocity of approach or recession of the gas cloud in question. Thus not only do we have here a tool for detecting H II regions, but we are also in a position to measure the radial velocities of the gas clouds participating in the process.

It is likely that less than 10 percent of the interstellar gas in our Galaxy is in the ionized condition. The most common form of hydrogen is probably neutral atomic hydrogen. It is most fortunate that the very abundant neutral atomic hydrogen can be observed through the radiation that it emits at a wavelength of 21 centimeters. The first observational evidence for the presence of the 21-centimeter radio radiation came early in 1951, when Ewen and Purcell of Harvard detected radiation with a wavelength near 21 centimeters reaching us from some sections of the Milky Way. This discovery of the 21-centimeter radiation was made because Ewen and Purcell had followed up van de Hulst's suggestion that such radiation should be observable. Once the discovery had been announced, other radio astronomers promptly confirmed and extended it. Principally through the efforts of Dutch and Australian radio astronomers, more recently with strong support from the United States, it has been possible to map with high angular resolution the whole of the Milky Way as it appears in 21-centimeter radiation.

We noted in Chapter 7 that the 21-centimeter line results from a transition between the lowest levels of the neutral hydrogen atom. A fraction of the neutral hydrogen atoms in interstellar space are excited to the slightly higher energy level, mostly through collisions. The atom will stay for quite a while in this very slightly excited state and then return to the very lowest of the levels. The 21-centimeter line is produced as the result of this transition. It is basically a very sharp spectral line. Since we know its rest wavelength very precisely, we can use the shift between the observed wavelength and the rest wavelength as a precise measure of the radial velocity of the cloud of neutral atomic hydrogen that produces this feature. The simplest way to represent an observation is by pointing one's radio telescope toward a certain direction in the Milky Way and then slowly scanning in wavelength through the line. If the receiving apparatus is of good modern quality, it will be possible thus to obtain a 21-centimeter profile with a radial-velocity resolution of 2 kilometers per second or less.

We shall see in Chapter 10 that the 21-centimeter line of neutral atomic hydrogen provides us with a very powerful tool for the study of the spiral structure in our own and other galaxies. It now appears likely that most of the 21-centimeter radiation recorded with our radio telescopes originates in relatively cool interstellar regions. Average temperatures of the order of $120°K$, $153°$ below zero on the customary centigrade scale, will be found in most of the regions where the 21-centimeter line originates. However, this temperature is very much of the nature of an average. We shall see later on that inside dark nebulae, which are clouds of cosmic dust, much lower temperatures may prevail.

We noted earlier in the present chapter that during the 1930's the first molecules were detected in interstellar space. High-dispersion optical spectra reveal the presence of the telltale absorption lines attributed to simple diatomic molecules, such as CH, CH^+, and CN. In the early 1960's, radio astronomy entered the picture in a big way. First, in 1963, there came the discovery of radio lines attributed to the hydroxyl radical (OH). Three major discoveries followed in 1968–69: ammonia (NH_3), water vapor (H_2O), and formaldehyde (H_2CO). In 1970 a number of additional radio lines were discovered, most notably a carbon monoxide (CO) line. It is surprising that some very complex molecules, for example methyl alcohol (CH_3OH), are found to be present in interstellar space. Our present list of interstellar molecules contains about 30 varieties. We note that most of these molecules are found in regions of space where cosmic dust prevails. The Orion Nebula, with all its associated gas and dust, the gaseous clouds near the center of our Galaxy, and the quiescent smaller and larger clouds of cosmic dust within a few hundred light-years of the Sun have proved to be the favorite hunting grounds for molecular gas. Molecules are continuing to pop up in surprising concentrations in many unexpected spots in the Galaxy. Some of these molecules are organic species with as many as seven atoms to a molecule. Carbon monoxide, formaldehyde, and the OH radical are commonly present. Molecules are often found in regions associated with strong infrared sources.

The hydrogen molecule, H_2, is probably the most prevalent type of molecule in interstellar space. Theorists have proposed mechanisms that would make it very abundant inside dark nebulae, which are composed of cosmic grains, and possibly also in dust-embedded objects that shine in the far infrared. The Princeton results obtained with the aid of the Copernican Satellite, show that molecular hydrogen is present wherever cosmic dust is found in abundance.

Galactic Cosmic Rays

Cosmic rays are charged particles that move at terrific speeds through our Galaxy. They are mostly nuclei of standard chemical elements probably produced by supernova outbursts and guided in their galactic paths by the weak magnetic fields that prevail in the Galaxy. They are an integral part of the interstellar medium, and they harbor a considerable fraction of the total energy available in the Galaxy. When we trace a galactic cosmic ray by recording its path in a specially prepared thick photographic emulsion, we really record the capture of a particle reaching us from interstellar space. The cosmic rays are for the present our only direct contact with particles that are known to have come from beyond the solar system. For this reason alone, they deserve careful study.

When cosmic rays manage to reach the Earth after penetrating our atmosphere, they have already been thoroughly disturbed in their motions by the Earth's magnetic field and by possible interplanetary fields. They have also been affected by the action of the solar wind, the stream of particles that is thrown out into space by the Sun's atmosphere. Cosmic rays were first detected about 60 years ago through the ionizing effects they produce in gaseous ionization chambers. In-formation regarding the directions from which they came could be obtained by tracing the continuing effects of one single charged particle on a string of ionization chambers, properly aligned. It was soon found that the Earth's atmosphere greatly affects all but the most energetic particles and that secondary cosmic-ray "showers" are observed on Earth as a consequence of highly energetic cosmic-ray particles interacting with atoms high in our atmosphere.

To learn more about cosmic-ray particles, observers traveled to the high mountains of the Andes and the Himalayas, and established cosmic-ray observatories in the American Rockies and elsewhere. Particle counters were sent up by balloons to heights of 100,000 feet and more, and observations were made in deep mines at high altitudes (like the copper mines of Peru) and from the bottoms of deep lakes, also at high altitudes. These researches made possible the study of the properties of the charged particles that compose the cosmic rays. The most abundant components were readily identified; they are the nuclei of the hydrogen atom, the proton, and of the helium atom, the alpha particle, which consists of two protons and two neutrons. But it soon became evident that nuclei of heavier elements, certainly including the iron atom with atomic number $Z = 26$, were present as well.

In the past two decades, technological improvements have assisted greatly in the further development of the observational study of cosmic rays. To the rather meager tool chests of the early cosmic-ray astrophysicists were added the thick nuclear emulsions, developed for the study of charged particles in cyclotron and synchrotron atomic research. Also, it has become possible to make use of man-made satellite vehicles to carry detection equipment beyond the atmosphere of the Earth and into interplanetary space; the Moon may provide a good base for future cosmic-ray research. The detection equipment itself is being constantly improved. Cosmic-ray counters are being made increasingly more sophisticated and new "track-etching" techniques have been applied to meteorites, with the result that elements heavier than iron have been detected. The most massive detected nucleus is now one of atomic number $Z = 106$, truly a transuranian nucleus!

Where do the cosmic rays originate—in our Sun, in our Galaxy, or beyond? The answer to these questions is that they are to some minor extent of solar origin, mostly at the lower energy levels, but that the major contributions, especially at the higher energies, come from our Galaxy—with possibly some contributions from other galaxies as well. In this book we are of course especially interested in cosmic rays originating in our Galaxy. As of now, supernova explosions seem to be the most likely source for cosmic rays originating within the Galaxy.

For the galactically interested astrophysicist, the Sun—as Peter Meyer of the University of Chicago has put it so well—is really a nuisance. Not only does the Sun produce some cosmic rays of its own, especially at times when solar flares erupt, but, most of all, the gentle but persistent solar wind of low-energy atomic particles emanating from the Sun disturbs and modifies the cosmic-ray flux through interactions between cosmic-ray particles and atomic nuclei of the solar wind. Small wonder that the cosmic-ray astrophysicist has learned that the best time for the observations of galactic and possibly of extragalactic cosmic rays is at the time of minimal solar activity, that is, near the times of sunspot minimum.

The Earth and its atmosphere further complicate the life of the cosmic-ray astrophysicist. The magnetic field of the Earth makes the cosmic rays that penetrate through it deviate from their original paths. The Earth's magnetic field affects the cosmic-ray particles to such an extent that for all except the most energetic particles it becomes very difficult to trace the original directions in which they moved before entering the Earth's magnetosphere. Moreover, interaction between cosmic-ray particles and the gases in the upper atmosphere produce secondary effects in the form of showers of ionized particles. The Earth's magnetic field and its atmosphere surely act as a shield against cosmic rays! Satellite research is assisting terrifically in the study of cosmic rays before they enter the Earth's atmosphere and are affected by its magnetic field. Ultimately it will become exceedingly important to carry satellite research beyond the inner parts of our own solar system.

Protons and alpha particles are, as we noted, the principal constituents of cosmic rays. Next in line come the elements with atomic numbers $Z = 30$ and greater, especially the iron group. The most abundant elements of low atomic weight are listed in Table 5, with their approximate cosmic-ray abundances indicated relative to silicon = 1.0. For comparison, we list beside these numbers the cosmic abundances of these same elements according to Cameron, who bases his values mostly on meteoritic abundances. Two basic results stand out from this tabulation. First, for most of the elements from carbon to iron the relative abundances of the atomic nuclei comprising the cosmic rays are remarkably similar to those of the elements in interstellar space and in meteoritic material. Second, cosmic rays contain excessive num-

Table 5. Abundances of elements in cosmic rays.

Chemical element	Atomic number (Z)	Cosmic ray abundance (Meyer)	Cosmic abundance (Cameron)
Lithium	3	1.2	0.000045
Beryllium	4	0.8	0.00007
Boron	5	2	0.00006
Carbon	6	7	13
Nitrogen	7	2	2.4
Oxygen	8	5	24
Neon	10	1	2.4
Sodium	11	0.4	0.06
Magnesium	12	1.2	1
Aluminum	13	0.2	0.1
Silicon	14	1	1
Argon	18	0.4	0.2
Chromium	24	0.6	0.01
Manganese	25	0.6	0.01
Iron	26	0.9	0.9

bers of atomic nuclei of the light elements lithium, beryllium, and boron as compared with the cosmic abundance of these elements.

It is interesting to note that there are electrons and protons among the cosmic-ray particles. It is not an easy matter to separate the truly cosmic electrons from those produced by the solar wind and by secondary effects in the Earth's atmosphere. Satellite observations near sunspot minimum give the best data relating to free electrons in interstellar and interplanetary space. We should add that the deuterium nucleus (heavy hydrogen, D) and the isotope of the alpha particle, He^3, have both been detected in cosmic rays by Fan and his colleagues, then at the University of Chicago. The abundance of deuterium appears to be less than 10 percent of that of He^3. These rather unstable elements are not likely to have been produced in the violent events that

are probably involved in the original production of high-energy cosmic-ray particles. They must almost certainly have come about as a by-product of interaction between original cosmic-ray particles and the interstellar gas. They could not possibly have survived travel through the vast distances between galaxies. In themselves, they are strongly suggestive of a galactic origin of cosmic rays.

In recent years, there has been much controversy among the experts in the field about the origin—galactic or extragalactic—of cosmic rays. On the whole it seems that the proponents of the largely galactic origin are coming out ahead. The suggestion that has received most attention is one made by Ginzburg and Syrovetski, supported by Shklovsky, all three of the U.S.S.R., which places the origin of cosmic rays in supernova explosions within the Galaxy. In a galaxy like ours there are approximately two or three supernova explosions per century. The amount of energy produced in one of these explosions is enormous and the fact that radio-synchrotron radiation is observed as coming from the direction of known supernovae remnants, such as the Crab Nebula, indicates that large-scale magnetic fields are associated with them. The atomic nuclei that are being thrown into space as by-products of supernova explosions will be accelerated by these magnetic fields and the high energies of the cosmic-ray particles can thus be understood. Model calculations by Colgate and White and others confirm that the observed high energies are quite reasonable. There is a continuing supply of supernovae in the Galaxy. This is just as well, for the heavy nuclei and the light ones like deuterium and the isotope He^3 will not likely survive in the Galaxy for more than a few million years. One would not expect to find these varieties of cosmic-ray particles if the

83. The Crab Nebula in Taurus (Messier 1). The photograph was made in red (H-alpha) light with the 200-inch Hale reflector. It represents the remnants of a supernova observed by the Chinese in 1054. The Crab Nebula is probably a continuing source of cosmic rays. (Courtesy of Hale Observatories.)

84. A section of the Gum Nebula. The filaments shown in this photograph are probably the after-effects of a supernova explosion that may have taken place as long as 20,000 years ago. The whole of the Gum Nebula is shown in the lower panel to the right of center in Fig. 3. (From a photograph made with the Curtis-Schmidt telescope of Cerro Tololo Inter-American Observatory.)

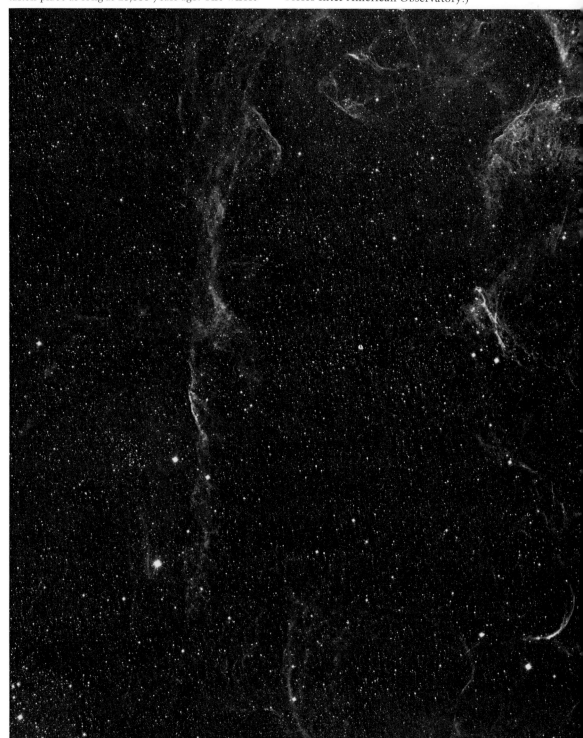

origin of most cosmic rays were extragalactic. The cosmic rays are certainly not coming to us to any great extent from galaxies at distances of the order of a few billion parsecs. The supernova hypothesis produces a steady supply of particles with just about the required amount of energy. It seems quite reasonable to look for the origin of the cosmic rays in the most spectacular events known to occur in our Galaxy, the supernova explosions.

Our principal concern about the hypothesis of the galactic origin relates to the fact that the highest-energy particles seem to come from all directions. If they were of galactic origin, then one would expect to have a preponderance of particles reaching us from directions pointing along the band of the Milky Way. To date there is no good evidence to support anything except isotropy of directions of arrival of the highest-energy particles; these are the particles most likely to arrive on Earth undiverted from their original paths. However, the story is by no means complete. A preliminary result of an analysis by Grote Reber, one of the originators of radio astronomy, indicates a considerably greater shower density coming from a band 20° wide centered upon the Milky Way than from the direction of a comparable area of the sky near the galactic poles.

From the study of cosmic rays we are already learning much that is new and interesting regarding the relative distributions of chemical elements and their cosmic abundances in galactic interstellar space. Cosmic rays give further proof for the presence of large-scale magnetic fields in the Galaxy. Cosmic rays are like tiny billiard balls that actually traverse the vast spaces between the stars of the Galaxy and, during their far-flung travels, interact with and suffer from encounters with interstellar matter. They arrive on Earth influenced by their experiences; they are ready to tell us about their adventures. As the late Arthur H. Compton, one of the early giants of cosmic-ray research, used to say, it is up to each of us to learn to read their messages from interstellar space. They are the only true galactic sidereal messengers.

During the past half century astronomers have increasingly paid more attention to the interstellar gas. In the first edition of this book (1941), we could devote a brief chapter to the interstellar gas and cover the field pretty well, but with each revision that chapter has grown in length. We have tried to present in the present chapter the most important facts about the interstellar gas and its composition, but our presentation is quite incomplete with regard to the physics of the gas. Some of these physical aspects are treated more fully in Chapters 7 and 12 of the Harvard Book on Astronomy, *Atoms, Stars, and Nebulae*, by Lawrence H. Aller (1971), but the time is obviously approaching when the interstellar gas and the cosmic dust deserve a Harvard Book of their own!

9
Dark Nebulae and Cosmic Grains

For the part of our Galaxy within 2,000 parsecs of the Sun and Earth, 80 to 85 percent of all matter is in the stars and the remainder, 15 to 20 percent, is in the gas and dust of interstellar space. According to the best current estimates, 99 percent of this interstellar matter is gaseous; tiny solid pellets account for the remaining 1 percent. In the present chapter we shall be concerned with observable effects produced by the interstellar dust particles and we shall see that much can be learned about their composition, approximate dimensions, and distribution inside and outside the cosmic clouds. Clouds of cosmic dust must reveal themselves in a variety of ways. Sometimes they shine like faintly luminous nebulae, but more often they appear as apparent star voids through which some distant stars may be seen. We can detect the presence and estimate the amount of intervening cosmic dust by the reddening and polarization effects in the light of remote stars.

Reflection Nebulae

Until 1912, it was generally supposed that all bright nebulae would show spectra similar to that of the Orion Nebula, which has a bright-line spectrum. Then V. M. Slipher announced that the nebula associated with the Pleiades has an absorption spectrum, very much like that of the brightest stars in the Pleiades cluster. Other nebulae were found to behave in a similar fashion. There are apparently two kinds of bright nebulae, one with bright-line (emission) spectra, the variety on which we wrote in the preceding chapter, the other with dark-line (absorption) spectra similar to those of the majority of the stars. Hubble showed that emission nebulae are generally near very hot stars with spectral types O, B0, or B1. He found that the nebulae with absorption spectra are associated with cooler stars. These nebulae are called *reflection nebulae*, since they shine by the reflected and scattered light from the stars that render them visible. They are truly starlit cosmic clouds. The reflection nebulae generally show no emission features because the stars that cause them to shine lack a sufficient supply of ultraviolet radiation to produce luminescence through ionization followed by recombination.

Struve, Greenstein, and Henyey made some careful early studies of the physical properties of reflection nebulae. From the observed surface brightnesses of these nebulae they deduced that the particles that produce the reflection are excellent reflectors. There is nothing gray or grimy about the dust of interstellar space. The cosmic grains resemble in their reflective power tiny hailstones much more than dust; in the language of the astronomer, the cosmic particles appear to have a high *albedo*, comparable to that of snow, though probably not quite so high. The indicated high reflectivity of these particles suggests that they are not primarily metallic. Most likely they are small icelike grains composed of simple molecular compounds of the lighter elements, such as carbon, nitrogen, oxygen, and of course hydrogen. Silicates too are identified in cosmic clouds; there is quite a bit of sandy stuff around!

Important conclusions regarding the probable dimensions of the particles may be drawn from comparative studies of the colors of these reflection nebulae and of the stars that illuminate them. We are all familiar with the observation that the light of the setting Sun is reddened considerably in its passage through the atmosphere and, in turn, that the scattered light from the Sun produces the blue aspect of our earthly sky. The contrast between the color of our Sun and the blue of our sky is explained by the phenomenon called *Rayleigh scattering*. The actual scattering of the Sun's light in our atmosphere is produced by particles with dimensions considerably smaller than the wavelength of visible light, probably mostly molecules. The reflection nebulae are somewhat bluer than the stars whose light they reflect, but the color difference is by no means so marked as in the case of our Sun and the daytime sky. From the observed differences in color between reflection nebulae and illuminating stars, we deduce that the scattering particles have average radii of the order of 0.00001 inch, one quarter of a micron.

Dark Nebulae

If a cloud of gas and dust is present in interstellar space, it will not appear as a diffuse nebula unless there is a bright star in or near it. Most clouds with cosmic dust do not shine, but such clouds will absorb and scatter the light from the stars beyond them and they will be seen as dark areas against the bright background of the Milky Way. We call them *dark nebulae*.

Dark nebulae are not conspicuous objects for visual observers. They are uninteresting regions devoid of stars, or with fewer stars than normal, and an observer will pass them by in favor of fields that are rich in stars. Sir William Herschel was the first astronomer who seriously considered the implications of the vacancies along the Milky Way, and it was also Herschel who noticed that these vacancies occur frequently in the vicinity of bright nebulae. It was not until a century after Herschel's observations that Barnard and Wolf succeeded in proving from their Milky Way photographs that many vacancies were caused by obscuring clouds rather than by real holes between the stars along the Milky Way.

We reproduce in this book several photographs of diffuse bright nebulae, emission nebulae as well as reflection nebulae. The many irregularities in the light reaching us from these bright nebulae, and the related irregularities in the distribution of the faint stars, suggest that many of the bright nebulae serve as backgrounds against which can be observed the numerous dark nebulae of various sizes that lie between these bright nebulae and our Earth. Among the finest examples are the Great Nebula near Eta Carinae (frontispiece) and the emission nebulae known as Messier 8 and Trifid (Figs. 85, 86, and 87). Lanes of obscuring material obviously overlie much of the bright structure, and we observe numerous small dark spots seen projected against the bright background. These spots are seen in the same positions from night to night and from year to year and they must surely indicate the presence of small obscuring clouds. Some of these spots have a wind-blown, turbulent appearance, whereas others—generally referred to as *globules*—have a markedly round appearance (Figs. 88, 89, and 90).

There are many known instances of association between bright and dark nebulae; the Horsehead Nebula in Orion is a good example (Fig. 49). The ectoplasmic glow around the horse's head emanates from the bright nebulosity. The horse's head is a part of the large dark nebula that covers most of the lower half of Fig. 49. If the dark nebula suggests an ominous thundercloud, then the bright nebula is the sunlit edge. The photograph of the Horsehead Nebula shows clearly the power of a dark nebula to absorb the light of the stars beyond it. If we compare the number of stars in two squares of equal size, one above and the other below the horsehead, we count at least ten times as many stars in the first square as in the second square. One can

have little doubt about the power of this particular dark nebula to dim the light of the stars that lie behind it.

The Coalsack (Fig. 91), in the Southern Hemisphere, is one of the most striking dark nebulae in a region devoid of bright nebulosity. Largely because of the contrast with the brilliant Milky Way surrounding it, the Coalsack appears to the visual observer as an intensely black cloud. Telescopic observations readily reveal the presence of numerous faint stars in the apparent inky blackness of the Coalsack. Long-exposure photographs show that there are still on the average one-third as many faint stars in the Coalsack as in an adjacent clear area of the same size. It is not really so black as it appears at first. The Coalsack Nebula stood for many years as the finest example of a dark nebula free from bright nebulosity. In 1938 a minute patch of bright nebulosity in the Coalsack was found by Lindsay at the Boyden Station, and others have since been noted by Gum; some of these are probably distant gas clouds viewed through the Coalsack.

Figure 92 shows the North America Nebula, so named by Wolf of Heidelberg. The photograph gives a striking illustration of the association between bright and dark nebulosity. The "United States" is a conspicuous bright nebula and the "Gulf of Mexico" is one of the densest portions of the surrounding dark nebula. From the large numbers of faint stars that shine through the bright nebula, it is apparent that the bright nebulosity is more transparent than the dark portions.

One of the finest early photographs of a dark nebula, made by Barnard, is reproduced in Fig. 93. The dark nebula near the star Rho Ophiuchi is a part of the giant dark-nebula complex in Ophiuchus, which, according to

85. Messier 8 and the Trifid Nebula. Two of the
finest emission nebulae north of the galactic center: Trifid near the top, Messier 8 (large) below.
Note the overlying patterns of dark matter.
(Uppsala-Schmidt photograph.)

86. The emission nebula Messier 8. A copy of one of the first photographs made with the Mayall reflector of Kitt Peak National Observatory. Messier 8 is the same emission nebula shown in the lower half of Fig. 85. Several dark markings are of roughly circular shape and these are referred to as "globules." (Photograph by D. L. Crawford.)

87. The Trifid Nebula in Sagittarius. (Photographed by N. U. Mayall with the Lick Observatory 120-inch reflector.)

88. The Rosette Nebula. A photograph in red light
of the Rosette Nebula in Monoceros, made with
the 48-inch Palomar Schmidt telescope.

89. Small globules of the Southern Milky Way. Thackeray at the Radcliffe Observatory first noted (1951) the remarkable dark markings shown here, which represent a variety of small globules. (Photograph on a red-sensitive emulsion made with the Curtis-Schmidt telescope at Cerro Tololo Inter-American Observatory.)

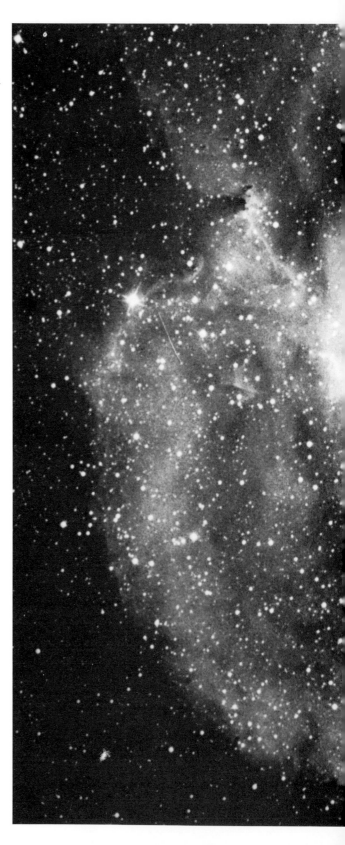

90. Globules in the Rosette Nebula. An enlarged
section of Fig. 88, showing numerous globules,
first noted here by Minkowski. (Hale Observatories photograph.)

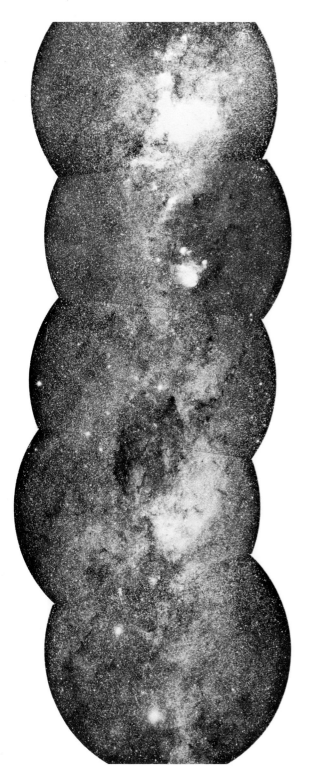

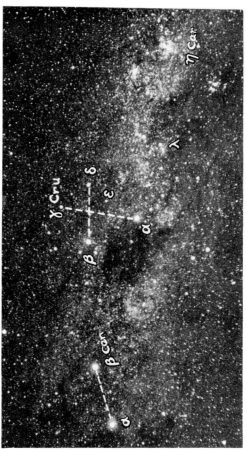

91. The Coalsack and the Southern Cross. The photograph at the top is a composite of five photographs taken in H-alpha (red) light that served later for reproduction in the Mount Stromlo Observatory *Atlas* by A. W. Rodgers, J. B. Whiteoak, *et al.*; see Fig. 3. The bottom photograph was made by A. R. Hogg with a small camera. The Southern Cross is marked and also the Pointers Alpha and Beta Centauri. The Eta Carinae Nebula is shown in the lower right-hand corner.

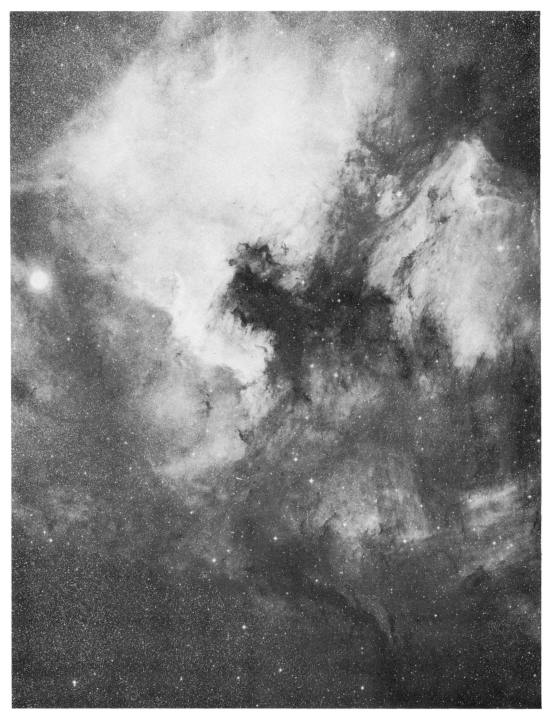

92. The North America Nebula and the Pelican
Nebula. Compare with Fig. 74. (Copyright by the
National Geographic Society–Palomar Observa-
tory Sky Survey.)

93. The dark nebula near Rho Ophiuchi. The obscuration in the center of dark nebula amounts to approximately 30 magnitudes of dimming. Near the edge of the pronounced obscuration, where two distant globular clusters can be seen on the photograph—a small one below the center and a larger one to the right of it and below—the obscuration amounts to only about 2 magnitudes. Most of the bright nebulosity is of the reflection variety. The cosmic dust is strongly concentrated at a distance of 200 parsecs. (From a photograph by Barnard at the Mount Wilson Observatory made with the 10-inch Bruce camera of the Yerkes Observatory.)

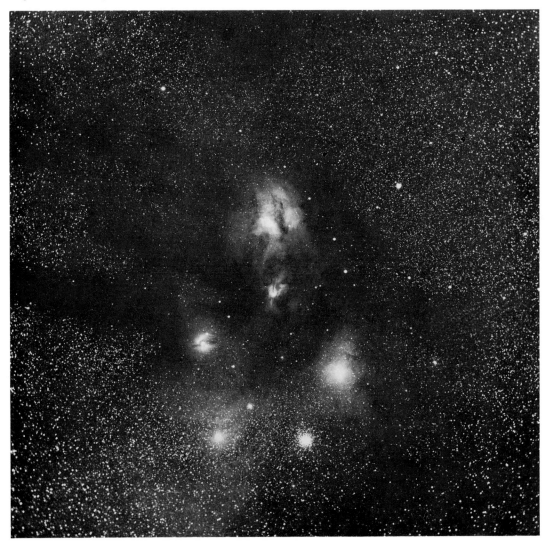

the evidence from the distribution of stars
and faint galaxies, covers an area of 1,000
square degrees. The great body of the nebular
mass lies from 5° to 20° north of the galactic
circle and lacks the background of faint stars
to render it conspicuous. The low-latitude
"tentacles" of the Ophiuchus dark nebula are
seen projected against some of the richest portions of the Milky Way and the dark nebula
near Rho Ophiuchus is one of these tentacles.
Other examples of Barnard's photography are
shown in Figs. 7 and 8.

Table 6 gives some indication of the types
of objects that concern us. Though there is a
terrific variety of dark nebulae, they can be
grouped roughly according to the three cate-
gories given in the table. The largest dark
clouds of cosmic dust that we observe with
well-defined boundaries are listed in the first
line of the table. We can estimate rather
closely the distances to these clouds. Clouds
of cosmic dust not only produce a dimming
in the light reaching us from stars beyond
them, but also have the effect of reddening
this starlight. Hence it is not difficult to distinguish between foreground and background
stars, and on the basis of such observations
we can assign rough distances to the various
clouds. From their apparent diameters in the
sky, we deduce their linear diameters. The
average radii given in the second column of
the table are thus derived. The typical large
cloud has a radius of about 4 parsecs and the
total mass of its cosmic dust alone is equivalent to about 20 solar masses. This mass estimate is based upon average values of the total

94. Varieties of dark nebulae. The section from a
red plate made with the 48-inch Palomar Schmidt
telescope (Sky Survey) shows the intricate dark
patterns observed against the background of the
star clouds in Sagittarius; see also Fig. 8. (Copyright by the National Geographic Society–Palomar Observatory Sky Survey.)

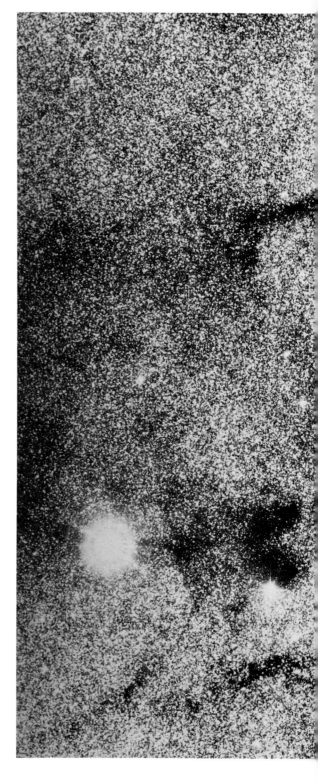

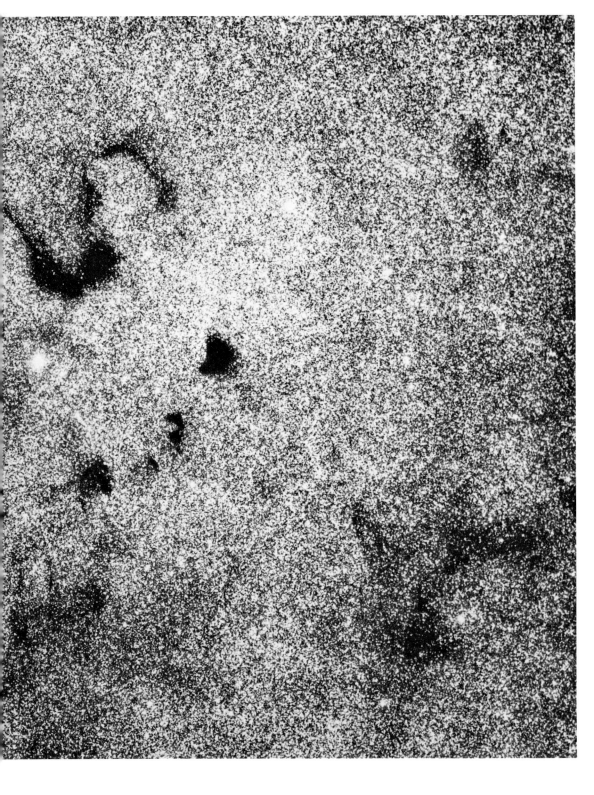

Table 6. Visible dark nebulae.

Object	Average radius (parsec)	Estimated mass (solar mass)	Accretion in 100 million years (solar mass)
Large cloud	4	2,000	1,000
Large globule	1	60	30
Small globule	0.05	0.2	0.05

absorption produced by the dark nebula and the known scattering properties of the small particles that produce the absorption in the first place. Various studies have suggested that there is much gas associated with these objects; the amount of gas is generally estimated to exceed the mass of the cosmic dust alone by a factor somewhere between 50 and 100. The estimated mass in the third column of the table includes the contribution from the associated gas. In the course of time, these clouds must sweep up considerable amounts of matter from the surrounding interstellar medium. It looks as if an amount of interstellar gas and cosmic dust equal to about half the estimated mass of the cloud is swept up in a time interval of the order of 100 million years. Roughly speaking, most of the well-delineated dark clouds should double their masses in time intervals of the order of 1 cosmic year.

In a way the globules, large and small, are the most interesting objects. The large globules often look on our photographs like "holes in heaven." In a region with a perfectly normal rich and smooth background of stellar distribution one suddenly encounters a darkened spot, which looks like an area of low sensitivity in the photographic emulsion. Longest-exposure photographs with large modern telescopes sometimes show the background stars faintly coming through the ob-

scuring matter associated with the globule. There can be little doubt that these dark holes are roundish clouds of cosmic dust floating by themselves in interstellar space. The table shows their estimated properties.

Finally, there are the small globules. They are seen most often projected as tiny dark specks against the luminous background provided by a bright nebula. We see no background light shining through them and the assigned masses in the third line of the table are only guesses at minimum masses. It has been impossible as yet to actually measure the gas content of globules. They have associated with them such small amounts of interstellar gas, especially of interstellar molecules, that observational evidence of the presence of gas is difficult to obtain. We should mention here that radio astronomers have discovered small and very dense units that contain the hydroxyl radical, OH; these are mostly near emission nebulae, but not at the positions of the small globules.

It is really very encouraging that we observe in space two varieties of dark globules and an assortment of large dark clouds. Their numbers are considerable. Within 1,000 light-years of the Sun, a very small distance compared with the diameter of our Milky Way System, we find about a dozen large clouds and about 100 fair-sized globules. We are not at all certain how many small globules there are. Small globules can only be seen projected against the bright emission nebulae, for they would cover areas of the sky too small to be distinguishable against stellar backgrounds. We therefore cannot say at present whether the small globules are selectively associated with the peripheries of emission nebulae, or whether they occur with reasonable regularity in our part of the Galaxy. On the whole, we would favor the first suggestion, since it is quite striking that small globules are absent from some emission nebulae. The small globules are probably clouds of dust and gas literally rolled up as little dust balls by the pressure exerted by the expanding gas at the periphery of the nebula. Their formation is probably assisted by the pressure of the ultraviolet radiation emitted by the hot O and B stars that are at the heart of each emission nebula.

The smallest globules measure only 0.05 parsec across (only 10,000 astronomical units) and the expected hydrogen density is close to 100,000 times that of hydrogen in interstellar space. They seem in many ways like protostars—stars and solar systems in the making. The larger globules are more massive and of course have larger diameters than their smaller equivalents, but they still seem close to the prestellar stage.

Clouds and globules seem to be units that have no choice except to collapse gradually into protostars, or break up into clusters of protostars. Whereas the pressure waves emitted by the bright nebulae may contribute to the formation of the smallest globules, the large clouds and the large globules will probably develop rather quietly on their own by the simple process of gravitational collapse. We need not worry about the rather high masses assigned to the larger clouds. Not every dark nebula needs to collapse into a single star. Instead, star birth may happen by simultaneous collapse of many subclouds in a Coalsack, or in a large-cloud complex. Tapia has found great irregularity in the distribution of the dark matter over the area of the Coalsack, and a number of globule-like units have been discovered.

How can we learn about the extent, distance, and composition of a dark nebula? Information on the first two points can be obtained from counts of stars in the area covered by the nebula and an adjacent area. The counts for two areas of similar size, one in the obscured region, the other in a neighboring apparently unobscured region, will generally agree for the brighter stars. As we count to fainter magnitudes, however, we soon come to a point where the counts in the obscured areas fall below those for the comparison area. The percentage deficiency will generally increase as fainter stars are included in our counts, but will finally assume a constant value. The apparent magnitude for which the deficiency begins to be noticeable gives us some idea about the approximate distance from the Sun to the dark nebula; the percentage deficiency for the faintest magnitudes is a good measure of the total obscuration caused by the cloud.

Wolf was one of the first astronomers to realize the value of star counts for the study of dark nebulae (Figs. 95, 96, and 97). Statistical methods for the analysis of such star counts were developed by Pannekoek. The large spread in the absolute magnitudes of the stars of all kinds renders it difficult to compute precise distances of dark nebulae, but Pannekoek's method of analysis tells us at least whether the absorbing cloud is at 100, 200, or 600 parsecs distance from the Sun. Pannekoek showed further that counts to faint limits give accurate and useful information on the total dimming produced by a dark nebula. The Coalsack, for instance, is caused by a dark cloud at roughly 170 parsecs from the Sun. That is right next door as galactic distances go. The total absorption of the cloud averages a little more than 1 magnitude, but in some dense portions is as high as 3 magnitudes; still larger values are obtained for Tapia's globules.

Estimated distances and absorptions are

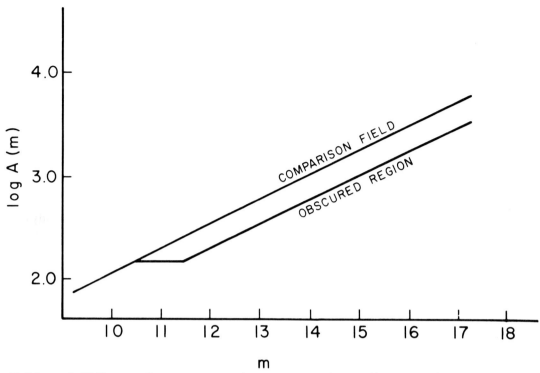

95. Schematic Wolf curves. Star counts are made for equal areas of sky in an obscured region and in a comparison field. The logarithms of the numbers of stars, $A(m)$, counted between apparent magnitude $m - \frac{1}{2}$ and $m + \frac{1}{2}$ are plotted (vertically) against the apparent magnitudes, m, shown horizontally. In the illustration the effect of the dark nebula is first noted near $m = 10.5$.

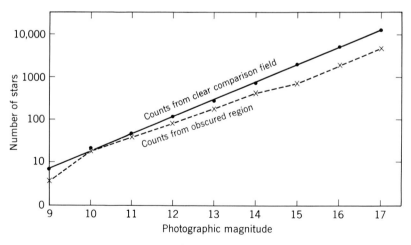

96. Wolf curves for a dark nebula in Cygnus. Star counts by Franklin at Harvard have yielded these curves for a dark nebula seen superposed on the star cloud in Cygnus. The solid curve gives the counts for the obscured field. The quantities plotted vertically are the numbers of stars between photographic apparent magnitudes $m - \frac{1}{2}$ and $m + \frac{1}{2}$, reduced to an area of 1 square degree of the sky.

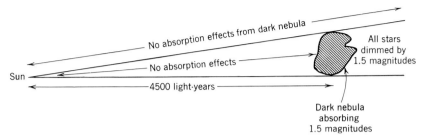

97. A model of a dark nebula. The star counts in the obscured field of Fig. 96 can be represented by assuming the presence of a dark nebula with a total photographic absorption of 1.5 magnitudes at a distance of 1,500 parsecs. (Franklin's results.)

now available for the dust clouds in most major centers of local obscuration. In the Rift of the Milky Way, the dark nebulae in Aquila and southward seem to be all relatively near the Sun, with estimated distances of the order of less than 200 parsecs; for the part of the Rift in Cygnus the distances are more of the order of 600 to 1,000 parsecs. The large dark-nebula complex in Taurus is again nearby, at 140 parsecs, but to the south we find in Orion, Monoceros, Puppis, and Vela a connected system of not-so-thick dark nebulae at 600 parsecs from the Sun.

General star counts alone tell us a good deal about the distance and absorption of a given dark nebula, but even with improved methods of analysis the results remain rather uncertain because of the great spreads in the absolute magnitudes of the stars. For definitive research on dark nebulae, we therefore turn to statistical studies based upon spectra and colors of faint stars. From the spectral statistics alone, we obtain an improved value for the distance of a dark nebula. The colors of the faint stars of known spectral type enable us to investigate in addition the extent to which the starlight has been reddened in its passage through the dark nebula.

The first extensive study of the effect of a dark nebula on the colors of faint stars ob-

served through it was made in the early 1930's by the Swedish astronomer Schalén, who found that although the starlight is definitely reddened in its passage through the nebula, the percentage reddening is rather small. The observed degree of reddening should depend almost entirely on the average dimensions of the cosmic dust particles in the nebula. Schalén concluded that the particles that are the most effective absorbers are tiny cosmic grains with diameters of about 0.00001 inch. Unfortunately, the observed coloring effects in the spectra of stars observed through dark nebulae do not permit clear-cut decisions about the composition of the particles. Schalén made the then rather reasonable assumption that the tiny dust specks are metallic, notably compounds of iron, zinc, or copper. Much has happened since the time when Schalén did his pioneer work. It has been proved that hydrogen and the lighter elements are much more abundant in space than the heavier metals. Also the reflectivity studies of bright reflection nebulae, about which we spoke in the preceding section, suggest that many interstellar particles are probably composed of carbon, nitrogen, oxygen, and hydrogen. Most astrophysicists are now inclined to favor the hypothesis of a nonmetallic composition for the cosmic grains, though serious doubts have again arisen since the discovery of interstellar polarization, about which we shall speak later in this chapter. Infrared studies are contributing much new information about the nature and origin of particles. Some stars that shine with unusual brightness in the infrared apparently possess very extensive dusty atmospheres. It looks as if silicate particles are being formed in these atmospheres. These particles are then probably expelled into interstellar space. Silicate particles do seem to represent a common variety of interstellar dust particles.

Ratio of Gas to Dust

The coexistence of interstellar gas and dust in many sections of the Milky Way has led astrophysicists to ask whether a constant ratio is maintained between the average densities of gas and dust everywhere in space. Attempts have been made to answer this question either through purely theoretical deductions or by trying to secure critical optical or radio observations.

The purely theoretical approach has been explored most thoroughly by Spitzer and Savedoff at the Princeton Observatory. If the gas component has a total density equal to at least 10 and probably 100 times the dust component, then it can be shown that the gas will in all probability literally drag the tiny dust particles along with it. The dust becomes little more than a tracer for the gas and a fairly constant ratio between the densities of interstellar gas and dust might be maintained everywhere. One recognizes, however, that this mixing in constant proportions may extend only over fairly small volumes and that there are many factors that may produce a different mixture for two widely separated points in our Milky Way system. Theoretical studies are helpful as a guide to the interpretation of observations, but theory alone cannot give a solution to the problem.

Unfortunately, the second approach, that of observation with the aid of either photographic and spectroscopic or radio techniques, has not provided a final clear-cut answer. Observations with conventional telescopes yield no clue, for the simple reason that the neutral hydrogen atoms in the cold interstellar clouds send out no detectable radiation in the photographic, visual, or red range. The neutral hydrogen will betray its presence only by the radio radiation with a wavelength of 21 centimeters.

Early studies of the radio radiation from neutral hydrogen were made by Lilley and others with the radio telescope at Harvard's Agassiz Station. Lilley found that the large Taurus complex of dark nebulosity is a region for which the intensity of the 21-centimeter radiation is considerably greater than the average for its galactic latitude. Where there exists a very large complex of cosmic dust, there is also a greater than average concentration of interstellar hydrogen gas. Lilley has shown for the Taurus complex that the average density of the neutral hydrogen is about 100 times the average calculated mass density of the cosmic dust. Hence, there is a great deal of neutral hydrogen for each dust particle.

It seems probable that the hydrogen in the denser dust clouds is either mostly in molecular form, H_2, or frozen as an H_2 mantle onto the tiny dust particles. The latter suggestion has recently been receiving much support. The solid particles that are the mainstay of the smaller dust clouds must have very low temperatures. Wickramasinghe, Reddish, and others have suggested that solid H_2 condenses onto cosmic grains, which they consider to be composed of graphite. In dark nebulae such as the Coalsack or in the tiny globules, the temperatures of the cosmic grains may be as low as 3°K. The presence of OH—the hydroxyl radical—molecules in very small clouds at the outer rims of H II regions is further proof that molecular binding processes are at work. This suggests also that the absence of neutral atomic hydrogen does not necessarily prove that all hydrogen is absent from the object being scrutinized. The graphite cores postulated by Wickramasinghe and Reddish have radii of 0.05 micron and the radius of the hydrogen mantle may be four times as great.

We mentioned in the preceding chapter that the Princeton analysis of data obtained with the Copernican Satellite has shown that molecular hydrogen prevails in some parts of interstellar space. The telltale absorption lines produced by the molecule H_2 are strong in the ultraviolet spectra of highly reddened stars, whereas they are mostly absent from the spectra of unreddened stars. Since reddening is indicative of the presence of clouds of cosmic grains between the star and the Earth, we may conclude that molecular hydrogen prevails in cosmic dust clouds. The Princeton estimates suggest that more than one-tenth of the hydrogen associated with the cosmic dust clouds is in molecular form. The corresponding fraction is less than one part in a million for the average of the interstellar gas. The hydrogen molecule, H_2, can form and continue to exist only in the cold dense clouds of cosmic dust, where the molecule is protected from the disruptive glare of ultraviolet radiation produced elsewhere by O and B stars.

The Cosmic Haze

In our study of interstellar material we have paid much attention to the obscuring clouds. What about the regions where faint stars shine in great numbers, where there is no direct evidence for the presence of intervening dust? We have learned to expect along the Milky Way the telltale absorption lines from the interstellar gas. But the atoms of hydrogen, calcium, and sodium can only absorb light of very special wavelengths and the presence of the gas will not lead to any general scattering or reddening of the light of distant stars. If we find scattering or reddening effects for distant objects outside the obvious obscuring clouds, then we shall have to leave room on our census blanks for a general haze of cosmic dust.

If we photograph any region of the sky that is at least 25° away from the galactic belt, we generally find quite a few images of faint spiral or elliptical nebulae on our plate. We refer to these as "galaxies" because they are separate stellar systems, some of which are probably not unlike our own Milky Way system. All of these galaxies are at distances from us that place them far beyond the borders of our Milky Way system. They begin to appear on 1-hour photographs with a 10-inch refractor. A 3-hour exposure with a 16- or 24-inch telescope will often show 100 of these galaxies for some sections of the sky outside the band of the Milky Way. They appear in abundance on all regular photographs taken with large reflectors of regions outside of the Milky Way, regions with galactic latitudes 40° to 90°, north and south.

Photographs of regions in the Milky Way, even long exposures with the most powerful telescopes, may not bring out any of these galaxy images. What does this mean? Can it be that there are no faint galaxies in those directions, or are they cut off from our view by a general interstellar absorption? There can now be little doubt that general galactic haze is to blame for the absence of galaxies on our

Milky Way photographs. There are probably distant galaxies in all directions. Those along the galactic circle are hidden from our view by the interstellar material in and near the central plane of our own Galaxy.

Now that we recognize the presence of an all-pervading cosmic haze close to the central plane of the Milky Way, we should attempt to find out by how much on the average the light of a star in the Milky Way at a distance of, say, 1,000 parsecs is dimmed by the haze. To do so we study distant stars of known absolute magnitude and estimate their distances by two methods, one basically trigonometric and hence not affected by interstellar absorption, the other wholly founded on brightness measurements. The latter produces distance estimates that are too great because they contain the full effect of the extra dimming due to interstellar absorption. We find the amount of absorption in magnitudes between the Sun and Earth and the star (or cluster) by finding the amount of interstellar absorption required to bring the trigonometric and the photometric distances into agreement. Once we know the total amount of absorption and the trigonometric distance, we can obtain a value for the average amount of absorption in magnitudes per 1,000 parsecs for the direction of the star under investigation.

The O and B stars, separate, in open clusters, or in associations, are the prefered objects for the application of this method, but cepheid variable stars of long period also are of value for studies of this nature. Unfortunately, since we need stars or groups of stars with distances of 1,000 parsecs or over, trigonometric parallaxes and proper motions are of little use. Hence we turn to the galactic-rotation effects in the radial velocities of O and B stars or cepheids and use Camm's

method, which we described in Chapter 6, to fix the average trigonometric distance for a group of these stars that seem to be all about equally distant from the Sun. We note that the distances thus obtained are free from effects produced by the dimming of starlight through interstellar absorption and hence represent true average distances. Next, from the known average absolute magnitudes of the stars, estimated from the spectrum-luminosity classes for O and B stars or from the period-luminosity relation for cepheid variables, we find the average distances d, uncorrected for interstellar absorption, from the formula

$$5 \log d = (m - M) + 5,$$

where m is the apparent magnitude of the star and M its absolute magnitude. We then determine how much intervening absorption we must assume to bring each pair of derived distances into agreement.

Trumpler found indications of an average dimming in the photographic range of 1 magnitude at a distance of 1,500 parsecs from the Sun for stars along the galactic equator. Rather comparable values have been derived from studies of galactic-rotation effects in radial velocities of distant stars, like the cepheid variables; the first dependable absorption estimates from cepheids were obtained by Joy. The geometric methods give at best only averages of absorption. Although such averages hold considerable interest, they give us little or no information about the absorption characteristics for a given direction in the Milky Way, for, as we have already noted, irregularity is one of the primary characteristics of the distribution of the interstellar dust that makes up the cosmic haze.

For the study of the absorption along a given line of sight, we prefer to measure and analyze space reddening suffered by individual distant stars. O and B stars, separate or in clusters, or cepheid variables are the primary targets in such studies. One measures, by photoelectric techniques or otherwise, the color index $B - V$ for the star and then predicts from the spectrum-luminosity class for the O and B stars, or from the periods of light variation for the long-period cepheids, what the value $(B - V)_0$ of the intrinsic color index of the star really is. One will generally find that the observed value $B - V$ is larger than $(B - V)_0$, the difference being attributed to the reddening effects of the intervening cosmic dust. This difference is called the *color excess* exhibited by the star, or cluster, in question.

It is not difficult to prove that space reddening is a general phenomenon. Its effects are shown dramatically in Figs. 98 and 99. A distant open cluster, which is shown clearly in Fig. 98 by Westerlund's infrared photograph (on the right), is inconspicuous in the photograph in yellow light (in the middle) and completely hidden from view in the photograph in blue light (on the left). Figure 99 shows a reproduction of two photographs of a globular cluster by Walter Baade. The photograph on the left was made on a red-sensitive emulsion, the one on the right on a blue-sensitive emulsion. The exposure times were adjusted so that stars with color indices like that of our Sun would show images of roughly equal size. It is obvious from the appearance of the globular cluster and from the field stars that interstellar absorption affects the blue photograph (right) much more than the red photograph. In some cases the effects of space reddening can even be observed visually. Observers with large reflectors who do research on the colors of B stars naturally develop the habit of checking at the eyepiece

upon the identification of a given B star by noting the marked blue-white color of these stars as contrasted with the more yellow or reddish colors of other stars. But when we look at a faint B star in the section of the Milky Way that contains the galactic center, where space-reddening effects are quite marked, the faint B star may seem yellowish, sometimes even distinctly reddish.

How do we correct the distances of reddened stars for the effects of interstellar absorption? We shall learn in the section to follow that the ratio between an observed color excess and the corresponding total visual absorption is fairly constant. The ratio is determined by the size distribution and chemical composition of the interstellar particles that produce the absorption. If the color excess E_{B-V} is defined as

$$E_{B-V} = (B - V)_{obs} - (B - V)_0,$$

then the corresponding total visual absorption A_V, expressed in magnitudes, is

$$A_V = 3.2 E_{B-V}.$$

The precise value of the transformation factor, 3.2, can be determined by the methods explained in the next section. This relation appears to hold for all except a few directions in the sky. The true distance of the star is then found by the relation

$$5 \log d = (m - M) + 5 - A_V.$$

Space Reddening and Cosmic Grains

It is obviously critically important for research on our Galaxy that we should learn as much as possible about the properties of the cosmic grains. They make up the many varieties of dark nebulae, large and small, and they produce the general cosmic haze. First of all, we want to know all we can learn about

98. A reddened southern open cluster. Three photographs of the same field by Westerlund show the effects of space reddening. From left to right we have a photograph on a blue-sensitive emulsion, one on a yellow-sensitive emulsion, and finally one on an emulsion sensitive to the near infrared. (Uppsala-Schmidt photographs.)

cosmic grains for their own sake, and second, we hope to derive reliable information on the value of the coefficient that relates the total absorption to the observed color excess.

The most comprehensive first attempt to derive the law of space reddening from observation was made by Whitford in 1948. By comparing the variation in brightness distribution with wavelength as observed in some highly reddened O and early B stars with the predicted brightness distribution on the basis of their spectrum-luminosity classes, he determined by how much the light of the star is reddened at each wavelength. Since the observations indicate decreasing reddening as we proceed from the blue-violet through the red toward the infrared, he assumed that there would be essentially no reddening left in the

farthest accessible infrared. He then derived at each wavelength the total absorption A, expressed in conventional magnitude units, obtaining a value for each pair of wavelengths, for example those corresponding to B and V. He read from his curve the value of E_{B-V}, the color excess, which, again, can be expressed in units of magnitude. If A_V represents the total visual absorption, Whitford's basic result is that on the average—and with very small scatter—we have

$$\frac{A_V}{E_{B-V}} = 3.2.$$

Following Whitford's precepts, we can obtain the total visual absorption A_V affecting the light of a star at the standard visual wave-

length by multiplication of the observed color excess E_{B-V} by 3.2.

So far so good! But there were some thunderclouds on the horizon as early as 1948. Back in 1937, Baade and Minkowski had shown that a conversion factor 3.2 did not seem to apply to the early B stars in the Sword of Orion, and a region in Cygnus was similarly suspect. There were indications of the presence locally of conversion factors as great as 6.

The Whitford results apply to the interval in wavelength from 3,500 to about 21,000 angstroms; the Earth's atmosphere gives an automatic cut-off in the ultraviolet at 3,500 angstroms, and Whitford's equipment did not reach in the infrared beyond 21,000 angstroms. About 1961, H. L. Johnson started work further in the infrared, including observations at 50,000 and 102,000 angstroms (at wavelengths near 5 and 10 microns). He found indications that the value of the ratio given above, 3.2, is a minimum value, but that for some directions values as great as 6 or 7 may apply. At first, Johnson's results caused much consternation. Recent work in the infrared has shown that the apparent far-infrared excesses found by Johnson are mostly intrinsic to the stars observed by him and that the conversion factor 3.2 still holds for most directions in the Milky Way. Deviations from average conditions do prevail, however, in some sections of the Milky Way, notably the heart of the Orion Nebula and a region in Cygnus.

Space research is adding new information

99. Two photographs of the globular cluster NGC 6553. This heavily obscured globular cluster was photographed twice by Baade with the 100-inch Hooker Telescope on Mount Wilson. The photograph on the left is on a red-sensitive emulsion, the one on the right on a blue-sensitive emulsion. (Hale Observatories photograph.)

on the law of reddening in the ultraviolet. Boggess, Borgman, and Stecher were the first to extend the curve from 3,500 to 1,400 angstroms. They found indications of a maximum of absorbing power near a wavelength of 2,000 angstroms. More precise recent results have come from Orbiting Astronomical Observatory No. 2. Bless and Savage have published an extinction curve that has a distinct maximum near wavelength 2,175 angstroms. Their tabulation lists the ratio of the color excess $E_{\lambda-V}$ to E_{B-V}, where λ represents the wavelength in the ultraviolet. Their representative values are as follows:

λ (angstroms)	$E_{\lambda-V}/E_{B-V}$
4,400	1.0
3,300	1.9
2,500	4.1
2,175	6.8
2,000	5.6
1,500	4.3
1,250	6.0
1,100	8.4

The precise wavelength of the maximum near $\lambda = 2,175$ differs a bit from star to star and the observed ratios also vary slightly from star to star. The gradually increasing value of the ratio as we proceed toward shorter and shorter wavelengths is rather to be expected, but the origin of the bump near $\lambda = 2,175$ is still a puzzle.

Infrared researches are yielding further useful information about the interstellar medium. The spectra of cool and very red supergiants show evidence of an absorption feature near $\lambda = 97,000$ angstroms, the presence of which has been stressed especially by Woolf and Ney at the University of Minnesota. There is good support for the suggestion that solid silicate particles are ejected in large numbers by the atmospheres of cool supergiant stars. Evidence of such particles is found in the infrared spectrum of the Orion Nebula. There is an indication also of the occasional presence of water ice, which possibly may form a mantle around a silicate particle; however, regular ice seems quite rare. Laboratory studies by Huffman and Stapp at the University of Arizona support the hypothesis of the presence of silicate particles of rather limited ranges of diameters. The polarization data obtained by Zellner, at the University of Arizona, indicate that reflection nebulae show infrared polarizations that also suggest the presence of silicates. We should note that the silicate interstellar feature, though clearly marked, is not really a very strong one. Woolf has estimated that for every 50 magnitudes of visual absorption at 4,400 angstroms there is only 1 magnitude of infrared absorption at 97,000 angstroms!

What do all these results imply with regard to the sort of particles that we encounter in interstellar space, and their probable dimensions? We have already noted that silicate particles may well be the most abundant ones in the interstellar medium. But it seems likely that there are also present some carbon compounds and an admixture of metallic particles, notably iron particles. "Dirty-ice grains," simple frozen aggregates of carbon, nitrogen, and oxygen, combined chemically with hydrogen, have been suggested and there probably also exist some solid particles with icy mantles. The important thing to note is that more and more evidence seems to be accumulating to suggest that many interstellar particles are created first in the atmospheres of cool stars. We have already presented the case for the silicates. The case for carbon compounds being formed in atmospheres of certain cool stars is also a convincing one. The R

Corona Borealis stars are a peculiar variety of variable stars. Most of the time they are at constant brightness, but then they dim suddenly, an effect caused by changes on a grand scale in their atmospheres. Apparently very large numbers of solid carbon particles form without much warning. Upon ejection these become interstellar particles—and the star recovers, only to get ready for the next dimming! We note that this basic mechanism was suggested as early as 1938 by O'Keefe, then a Harvard graduate student.

The small particles ejected by cool stars should be fine centers for condensation once they are off by themselves in the interstellar medium. They are cold, with temperatures ranging from $100°K$ ($173°$ below zero centigrade) to $3°$ to $5°K$. We may expect that interstellar atoms which collide with the particles should frequently stick, literally freeze onto them. The prevalence of molecules in dark nebulae shows that molecule building takes place in interstellar space, and where could an atom find more hospitable surroundings to mate with another atom (or two) and form a molecule than at the surface of a cold interstellar particle inside a dark nebula? The dark nebula provides incidentally a fine shield to protect the young molecule from potentially destructive ultraviolet radiation.

Interstellar Polarization

Polaroid sunglasses take most of the glare out of the Sun's rays reflected by the surface of a roadway because they remove effectively the rays in the plane of vibration that is strongest in reflection. Effects of light polarization are observed in the scattered light from the corona of our Sun and in the atmospheres of planets, but until 1949 no one expected that the light of distant stars would become polarized as a result of its passage through the interstellar medium.

To produce interstellar polarization in the light of distant stars requires not only a preponderance of somewhat elongated particles in interstellar space, but, further, some powerful mechanism to align these particles over very great distances. This, so it was reasoned, would not likely occur and it therefore came as a great surprise when, in 1949, Hall at the U. S. Naval Observatory and Hiltner at the McDonald Observatory announced simultaneously that they had succeeded in detecting polarization effects in the light of distant stars of our Galaxy. The observations of Hall and Hiltner were made with photoelectric polarimeters. These are instruments that permit comparative measures of any star's brightness in different planes of vibration. The recording by photoelectric means guarantees a remarkably high precision of measurement, which is essential when one deals with observed effects as small as those of interstellar polarization. Even in the most extreme case, the differences in intensity for the planes of greatest and of least intensity amount to no more than 0.15 magnitude, and for the majority of the more distant reddened stars the differences are less than 0.03 magnitude.

A sufficient variety of distant stars has now been observed that we can say definitely that the polarization is produced by the solid particles of interstellar space. Polarization affects the light of all distant stars along the belt of the Milky Way, irrespective of spectral class or absolute magnitude. Strong polarization effects are observed only for stars that are highly reddened by intervening cosmic dust, even though strong reddening is not necessarily accompanied by high percentage polarization. The fact that polarization and reddening generally do go together speaks strongly in favor of the hypothesis that polarization is associated with cosmic dust.

This conclusion is strengthened by the further remarkable observation that similar polarization characteristics are often found over fairly large sections of the Milky Way. In the direction to the constellation Perseus, for example, distant stars over a large area of the sky show comparably high degrees of polarization and a close alignment of their principal planes of polarization. Such phenomena can be understood only if one assumes that the polarization is produced by interstellar clouds or cloud complexes, with dimensions of the order of several hundred parsecs, composed of elongated particles with roughly parallel alignments. In some sections of the Milky Way the parallel alignment is very marked, whereas in others the planes of polarization present a more jumbled picture. This is shown very nicely in Fig. 100.

To produce detectable polarization effects in the light of distant stars requires first of all that there be in interstellar space somewhat elongated particles and, second, that some mechanism exist to align these particles in a more or less parallel fashion over very large regions. There is now little doubt that the particles are aligned mostly by large-scale interstellar magnetic fields. To obtain the necessary degree of alignment for the explanation of the observed polarization effects requires at least egg-shaped particles, with some metals (iron?) present and, according to the best available theory, that of Davis and Greenstein, magnetic fields of the order of 0.00001 gauss. In this theory the tiny interstellar particles would be spinning rapidly. The assigned values of the magnetic fields are infinitesimal compared with those found for the Sun and the Earth. However, the effects of these extremely weak fields are felt over hundreds of parsecs.

At present, the Davis and Greenstein theory seems to meet with the most general approval. There are, as always, some remaining problems. The theory requires the presence of magnetic fields over large stretches with field strengths that are greater by a factor 10, more or less, than the strongest large-scale magnetic fields supported by radio-astronomical evidence, which indicates fields of only a few millionths of a gauss close to the galactic plane. B. T. Lynds and Wickramasinghe have suggested that graphite cores with ice mantles can most readily explain polarization effects as observed in the presence of magnetic fields with strengths in the range 0.000010 to 0.000003 gauss.

Very useful information regarding the properties of the polarizing particles can be derived from studies of the variations with wavelength of the amount of polarization shown in the highly polarized light of some stars. Gehrels and associates at the Lunar and Planetary Laboratory and Visvanathan at Mount Stromlo Observatory have especially studied these phenomena. In most stars the polarization is greatest between 5,100 and 5,800 angstroms, and the percentage polarization is found to decrease toward the red and the blue-violet. Balloon observations show different rates of dropping off of percentage polarization in the far ultraviolet. Some of these differences may have their origin in the polarization properties of stellar envelopes, rather than in the interstellar medium, a hypothesis advanced by Serkowski. However, the observed sharp decreases in percentage polarization toward the far ultraviolet seem to speak in favor of the presence of graphite grains with an ice mantle, which, according to calculations by Wickramasinghe and others, should show this sort of wavelength dependence.

One of the most remarkable phenomena suggesting the presence of large-scale magnetic fields is the alignment of wisps of dark and bright nebulosity for rather large sections

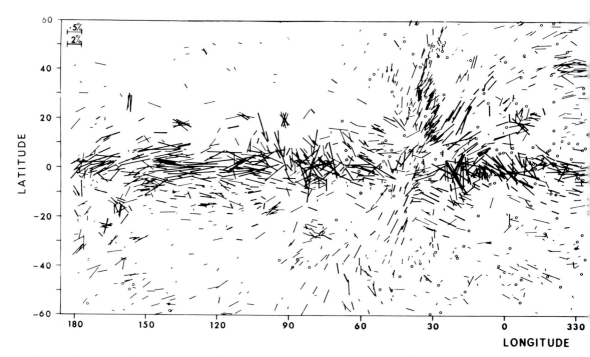

of the Milky Way; Shajn and Rouscol have especially drawn attention to this phenomenon. Shajn derived for some sections the direction of the magnetic field that he holds responsible for the observed elongations of the dark nebulosities. He found (see Fig. 101) that the orientation of this magnetic field agrees, within permissible limits of error, with that derived from the observed polarization effects for stars seen through these nebulae.

It should not be thought that polarization effects are limited to cosmic dust. Shklovsky predicted from theory that strong polarization effects might be expected from very fast-moving electrons in magnetic fields of interstellar space. Dombrovsky observed the effect in the Crab Nebula (Fig. 83); his measures were confirmed and extended by Oort and by Walraven, who found close to 100-percent polarization in certain parts of the Crab Nebula. Baade has made some very high-resolu-

tion photographs of the inner filamentary structure of the Crab Nebula and he finds filaments with diameters of the order of 1 to 3 seconds of arc that show 100-percent polarization. The polarizations produced by these so-called synchrotron-radiation effects are inherent to the object—the Crab Nebula in this case—that exhibits them. From such effects we gain no information about the properties of the interstellar medium.

Data on Space Reddening

The most complete picture of space reddening along the band of the Milky Way is obtained through measurements of color excesses of O and B stars. The first accurate photoelectric survey was completed by Stebbins, Huffer, and Whitford at the University of Wisconsin and the Mount Wilson Observatory (1939). Their basic list contains precise

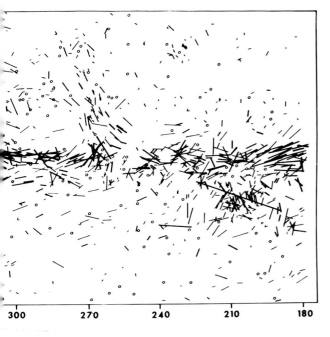

100. Optical polarization. This diagram summarizes all available data on interstellar galactic polarization. The plot shows the electric vectors of polarization and represents data for 7,000 stars. Measurements for 1,800 stars were made by Mathewson at Siding Spring Observatory; the remainder are from the catalogues by Hall, Hiltner, Behr, Lodén, Appenzeller, Visvanathan, and E. van P. Smith. Small circles: positions of stars with percentage polarization $P < 0.08$; thin lines: vectors for stars with $0.08 < P < 0.60$; thick lines: vectors for stars with $P > 0.60$. (From a paper by Don S. Mathewson in the *Memoirs of the Royal Astronomical Society*.)

101. Relation between dark nebulosities and polarization. Shajn calls attention to the tendency for long filamentary structures in emission nebulae and in dark clouds to lie parallel to the galactic equator. This figure shows a region in Perseus-Taurus with dark nebulosities. The short lines indicate the direction of the interstellar magnetic lines of force as deduced from the observations of the polarization of light for distant stars made by Hiltner and Hall.

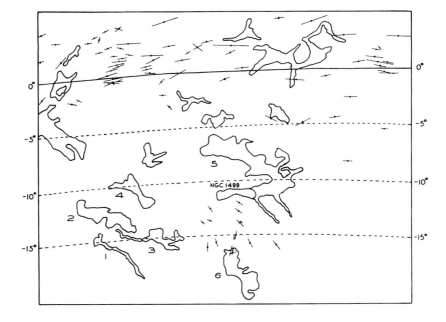

photoelectric colors for 1,332 O and B stars. Because of their great intrinsic brightnesses, even rather bright O and B stars are distant objects and they are therefore quite suitable for work on space reddening. The research has been extended and continued by many workers in the field in both Northern and Southern Hemispheres. For the majority of O and B stars we now possess precise determinations of color excess and of spectral absolute magnitudes on the Morgan-Keenan-Kellman system. Observed color indices on the *UBV* Johnson system and on the Strömgren *uvby* system are used for determinations of color excesses. A wealth of observational material on space reddening therefore is now available.

High reddening prevails in some sections of the Milky Way, in particular for the region of the galactic center. Bok and van Wijk found one faint B star in an obscured field not far from the direction of the galactic center which was so reddened that its estimated total photographic absorption is close to 6.5 magnitudes. In other words, this tenth-magnitude B star would be a fairly conspicuous star, photographically between the third and fourth magnitudes, if it were not for the obscuration between the star and our Sun. Space reddening is generally much less marked away from the direction of the galactic center, but to the best of our knowledge all faint O and B stars directly along the Milky Way show measurable effects of space reddening. This suggests that nowhere along the Milky Way is there a direction totally free from the general haze. The historic survey of Stebbins, Huffer, and Whitford showed that very highly reddened B stars are apparently absent between galactic longitudes 150° and 240°, the region away from the galactic center, but they are numerous for the region between galactic longitudes 350° and 30°, which includes the galactic center.

In recent years much attention has been given to color studies of distant cepheid variables along the Milky Way. At Mount Stromlo Observatory in Australia, Eggen and Gascoigne have studied the cepheids of our Galaxy and compared them with similar stars observed in the Magellanic Clouds. Gascoigne has made the rather startling suggestion that probably all the cepheids observed in the Galaxy are considerably reddened and that we observe unreddened cepheids only in the Magellanic Clouds. Now that we all agree, more or less, on the absolute magnitudes of the long-period cepheids, they become one of the most useful types of objects for the study of space reddening.

Faint and distant globular clusters have come in for their share of attention. Stebbins and Whitford measured photoelectric colors for all those within reach from the Mount Wilson Observatory, and Irwin has checked their survey and extended it to the Southern Hemisphere. Comparative studies of the color excesses for the very remote globular clusters in the direction of the galactic center, and for the relatively nearby B stars in the same general direction, have shown that the strong interstellar absorption is concentrated in the outer part of our Milky Way system, that is, in the region where the Sun and Earth are located. Even for the most transparent sections of the Milky Way in the direction of the galactic center, the indications are that two-thirds of the total observed obscuration occurs within 2,000 parsecs of the Sun, leaving the remaining 8,000 parsecs that separate us from the galactic center relatively free from cosmic dust. However, we do remember that the center itself is probably viewed through 27 magnitudes of visual absorption!

How regular or irregular is the distribution of the interstellar dust in the general haze? Is it smooth or does the superposition of many single dust clouds produce the total effect we observe? At present it is impossible to give a positive answer to these questions. There is apparently no region along the galactic circle where the view is perfectly clear. In a few places some galaxies shine through the haze in the galactic belt, but even there the number of faint galaxies is far below par and a total of several magnitudes of visual absorption is indicated.

Because of the irregularities in the total distribution of the faint external galaxies (they are inclined to come in bunches too!), estimates of total absorption from galaxy data for a given area may well be in error by as much as half a magnitude. In spite of such uncertainties, the observed deficiencies in the numbers of faint galaxies can tell us much about the extent of some of the largest single clouds. The large dark-nebula complexes in Orion and Taurus, in Cepheus, and in Ophiuchus are not only star-poor regions but also very deficient in faint galaxies. The strong absorption by the dark complex hides the galaxies in these directions from our view.

We should not underestimate the total effect of the isolated dark nebulae. Only the nearest of these nebulae will be discovered by an inspection of Milky Way photographs. The Southern Coalsack is conspicuous to us because it is within 200 parsecs and covers a large angular field. But we should realize that it would hardly have been discovered if it had been ten times as far away, at 2,000 parsecs. Not only would it then cover only 1 percent of its present area, but it would further lack contrast because of the many foreground stars. Greenstein has computed that the known dark nebulae alone may account for 30 percent of the total haze for distances up to 1,000 parsecs.

In the late 1930's a group of Soviet astronomers led by Ambartsumian introduced the hypothesis that the interstellar absorption may be produced entirely by chance agglomerations of small obscuring clouds, with average radii of the order of 8 parsecs, with the average absorption per cloud of the order of 0.2 magnitude, and with the average line of sight intersecting about one cloud every 20 parsecs. Several lines of evidence suggest that this is not the whole story. First, we note that the known dark nebulae are by no means distributed at random; they tend to congregate in large complexes, such as the Taurus and Ophiuchus dark nebulosities. Second, we find several large smooth stretches of the Milky Way where there is little or no indication of structure. In these sections, marked space reddening is present and nowhere is the total photographic absorption found to be less than half a magnitude at 1,500 parsecs from the Sun. Finally, we should mention that nowhere have we found any clear "holes"—patches of clear sky between the cumuli—which should occur under the pure cloud hypothesis. The current trend is then to look upon the interstellar absorbing medium as a turbulent affair. As far as we can tell today, there is probably some continuous haze, but it seems that the isolated dark dust clouds may be responsible for more than half of the observed absorption and scattering. We certainly have a cloud structure, with clouds of all sorts of dimensions contributing, and in addition there seems to be, close to the galactic plane, an all-pervading cosmic haze.

10
The Spiral Structure of the Galaxy

Our Galaxy consists basically of three parts: the nuclear region, extending to a distance of about 5,000 parsecs from our Sun; the thin disk, not more than 600 parsecs thick at the Sun, which contains the most spectacular Population I objects and in which spiral structure prevails; and the outer halo, mostly inhabited by Population II stars. In the present chapter we shall be exclusively concerned with the parts showing evidence of spiral structure. The spiral region extends in the galactic plane from about 5,000 to 15,000 parsecs from the center. We note that the Sun is placed rather centrally in this spiral belt at a distance of about 10,000 parsecs from the center.

History of Spiral Structure, 1949 to Present

In the late 1930's, when the authors wrote the first edition of this book, it seemed an almost hopeless task to undertake the tracing of the spiral structure of our Galaxy. A real breakthrough came in the late 1940's, when Baade and Mayall reported results of their studies relating to the spiral structure in the Andromeda galaxy, Messier 31. They found that the spiral arms in that galaxy were most clearly traced by emission nebulae and by cosmically young O and B stars. Clusters and associations of O and B stars were especially helpful in outlining the spiral structure. Baade called on astronomers to examine in our Galaxy the distribution of O and B associations and their related nebulosities. The challenge was accepted by W. W. Morgan of Yerkes Observatory, who, in 1951, with two of his young students, Osterbrock and Sharpless, presented the first over-all picture. They found three parallel sections of spiral arms clearly delineated. The first of these they called the *Orion Arm;* it is the arm in which they locate our Sun, near the inner edge. Next they traced a portion of the parallel outer arm, the *Perseus Arm*, about 2,000 parsecs farther away from the center of our Galaxy than the Orion Arm; third, they found evidence for a section of an inner arm, the *Sagittarius Arm*, about 2,000 parsecs closer to the center than the Orion Arm. Figure 103 shows the Morgan-Osterbrock-Sharpless diagram brought up to date; it includes data from both the Northern and the Southern Hemispheres. Based largely on researches by Wilhelm Becker and Fenkart of Basel and by Schmidt-Kaler of Bochum, the diagram shows the positions in the galactic plane of the clusters and associations with O to B2 stars and of the associated emission nebulae. The three sections of the spiral arms found by Morgan are shown rather neatly, and it is on the basis of this sort of diagram that Becker asserts that the spiral arms of our Galaxy have an average pitch angle close to 25°. The pitch angle is defined as the angle between the direction of a section of a spiral arm and the direction of circular motion.

Early in 1951 radio astronomy began to stir. Ewen and Purcell in the United States and Christiansen in Australia had followed up Van de Hulst's suggestion and discovered the 21-centimeter line of neutral atomic hydrogen. Thus they had found a way of pinpointing the hydrogen gas not only in the three local sections of spiral arms but also in very distant spiral features. Most of the latter are hidden from the optical astronomer's view by intervening thick cosmic dust clouds. The Dutch radio astronomers, led by Oort and van de Hulst, were the first to have published a radio 21-centimeter map of our Galaxy.

The basic analysis of observational material from a 21-centimeter-line survey of the distribution of neutral atomic hydrogen in the Galaxy follows a rather simple pattern. First we select the section of the Milky Way that we wish to study. For example, it may stretch from galactic longitude 275° to 305° and cover the zone in galactic latitude from −10° to +10°. We establish within this area a fine network of positions for which we desire basic information. Along the galactic equator we may place these positions 0.5° or 1° apart in galactic longitude, and we may have rows of additional positions, perhaps not quite so tightly spaced, at every degree of galactic latitude from −10° to +10°. To obtain relevant material for a network of points that is as tight as indicated, we should have access to a radio telescope with an angular resolution of about 0.5° or better. A radio telescope of 100-foot aperture with a surface precise to 1 centimeter serves very well for the purpose. We should use this instrument with receiver equipment of high frequency resolution, so that features in the line profile separated by a frequency difference corresponding to a radial velocity of 1 or 2 kilometers per second will stand out clearly as separate features. The observations at each position consist in pointing the radio telescope in the right direction, something that must be done with high precision, and then recording the profile of the 21-centimeter line in that direction. This is generally done with multichannel receiving equipment, in which each channel successively or simultaneously records the intensity of the radiation in a narrow band of frequencies. The observation for each position consists of a trace in which intensity of the hydrogen 21-centimeter radiation is plotted as a function of frequency; we call this a *21-centimeter profile*. Since for a cloud of cool hydrogen gas at rest the profile would be very narrow indeed, each frequency in the observed profile can be taken to correspond to a certain radial velocity of approach or recession of the hydrogen cloud. The rest frequency of the

21-centimeter hydrogen radiation is very precisely known and it is hence not difficult to translate each observed frequency into a radial velocity of approach or recession of the hydrogen cloud to which the observation refers. Hence the final data for an observation of each point in the network can be shown as a plot of intensity of 21-centimeter radiation versus radial velocity. Typical profiles are shown in Fig. 104.

Once the network of profiles has been obtained, the radio astronomer faces the difficult problem of assigning approximate distances to the various features in the profile that occur at certain radial velocities of approach or recession. In the original interpretation of such profiles by the radio astronomers in Leiden and in Sydney, the assumption was made that there exists a single well-defined rotation curve of circular velocities for our Galaxy. The standard curve of M. Schmidt is shown in Fig. 105. It is then possible to fix for any direction of galactic longitude precisely what would be the distance of a cloud that shows a certain radial velocity of recession or approach. This follows, of course, on the assumption that all hydrogen clouds move around the center of our Galaxy with the precise circular speeds assigned to them by the basic rotation curve.

However, the situation is not so simple. First of all, and not unexpectedly, the clouds show motions of their own, mostly of the order of ±6 to 10 kilometers per second. What is more disturbing, it soon became evident also that our Galaxy does not possess a single average rotation curve, in which the circular velocity of galactic rotation is only a function of the distance of the cloud from the center. It was at one time thought that there were two different kinds of rotation curves, one for the northern half of our

102. The southern Milky Way. The four photographs show nicely most of the southern Milky Way from Carina through Norma to Scutum. (*a*) The Eta Carinae Nebula and the Southern Coalsack above and slightly to the right of center, photographed in red light (wavelength 6,600 to 6,900 angstroms). The brightest area below the center is the Sagittarius Cloud. The bright lines on the right are from the seeing tower and from the building of the 152-centimeter reflector at the La Silla Station of the European Southern Observatory in Chile. (*b*) A photograph taken in visual light (wavelength 4,600 to 5,800 angstroms). The Southern Coalsack is just above and to the right of the shadow of the objective and the Sagittarius Cloud is to the left of the shadow of the lower strut. The shadows shown on the right are from the dome of the Danish telescope at La Silla and from the building of the 152-centimeter reflector. (*c*) A photograph taken in blue light (wavelength 3,900 to 4,700 angstroms). This photograph, like the others, shows especially well the dark constellation of the Emu to which reference is made in Chapter 1. The Southern Coalsack marks the sharp beak of the Emu (upper right). The long thin neck of the animal and the main body are seen directly to the right of the shadow of the objective and the legs are seen below the shadow of the right strut. The bright image to the left of the Milky Way and below the center is that of the planet Jupiter; the planet is not shown quite so conspicuously in the other photographs, but with a little effort it can be readily identified in each reproduction. (*d*) An ultraviolet photograph (wavelength 3,200 to 3,800 angstroms). The Emu seems to have more than two legs! (Photographs by Schlosser and Schmidt-Kaler made with the Bochum University wide-angle camera at the Bochum Station at La Silla in Chile.)

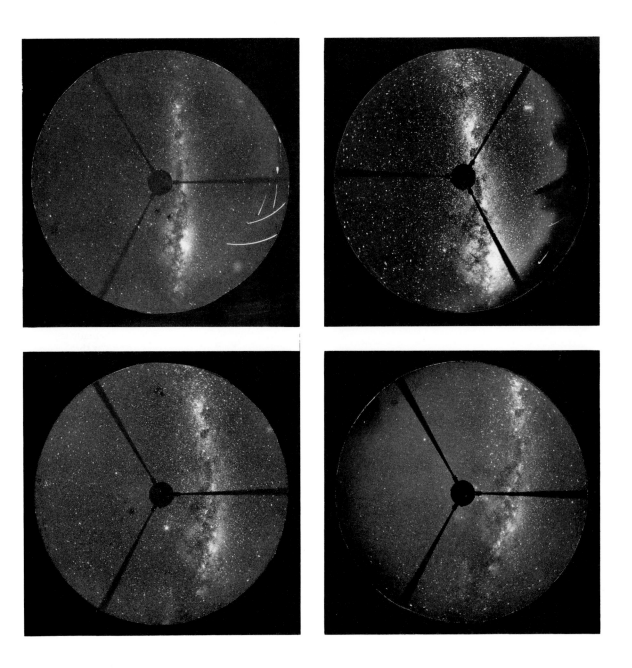

103. Optical spiral structure of our Galaxy. The Sun is at the center of the diagram. The galactic center is in the direction toward 0° galactic longitude at a distance of 10 kiloparsecs from the Sun. The principal observed sections of the Sagittarius, Orion, and Perseus Arms are shown. The directions from the Sun toward some of the key constellations along the band of the Milky Way are marked along the periphery of the diagram. (A diagram based on data from W. Becker and Th. Schmidt-Kaler.)

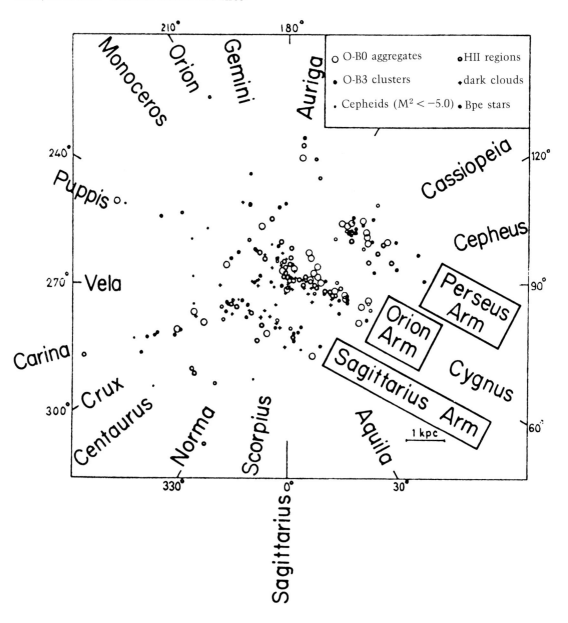

104. Typical 21-centimeter H I velocity profiles for adjacent centers in the southern Milky Way at two galactic longitudes, $l = 296°.5$ and $l = 297°.5$, for galactic latitude $b = 0°.0$. The abscissae give the radial velocities of the neutral hydrogen gas. The velocities have been corrected for the effects of the Sun's local motion. The ordinate gives the radio intensities of the signal in terms of the antenna temperature T_A in degrees Kelvin. If we assume that all the peaks are due to concentrations of H I and not to velocity crowding effects, we find for the H I clouds at $l = 296°.5$ the following radial velocities: -28, -7, $+15$ to $+20$, $+55$, and $+114$ kilometers per second.

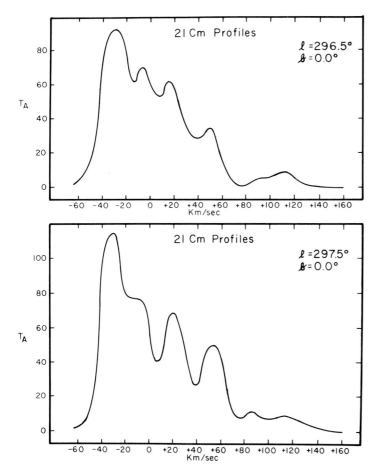

Milky Way band, and another for the southern half; Fig. 106 shows the two curves. This is probably not the case, but recent work has shown that there are large-scale velocity streamings present in the interstellar gas, which make it a difficult matter to transform the observed radial velocity of a cloud at a given galactic longitude into an estimated distance for that cloud.

The present state of interpretation of the large body of data on 21-centimeter profiles is rather confusing. A comprehensive review of the situation that we face was made at the 1969 Basel Symposium organized by the International Astronomical Union. We can do no better than to reproduce here two recent diagrams of radio spiral structure of our Galaxy—one by Kerr, the other by Weaver; these are shown in Figs. 107 and 108. The diagram presented by Kerr is based upon joint analysis by himself and Westerhout. It indicates a rather tightly wound pattern of spiral features, leading to spiral arms with average pitch angles of the order of 5°. The Weaver diagram shows a much more open spiral pattern, and he assigns pitch angles to sections of the arms in the range between 4° and 12°.

Before we attempt to reconcile Figs. 103, 107, and 108, we should have a look at some of our neighbor galaxies that obviously possess spiral structure. We obtain guidance in the analysis of our own Galaxy from the results of studies of these neighboring galaxies, for in them we can view the whole of the spiral pattern from a single photograph, or from a comprehensive analysis of 21-centimeter profiles. For our own Galaxy, we cannot hope to obtain such an over-all picture simply because the Sun and Earth are immersed in the Galaxy. To guide our attempts at analysis, we need the data from the Andromeda galaxy and from others, especially Messier 51.

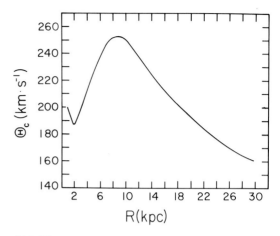

105. The rotation curve of our Galaxy. The dynamic model for our Galaxy by M. Schmidt yields the rotation curve shown in the diagram. Vertically the circular velocity Θ_c is given in kilometers per second, for distances (horizontally) from the galactic center measured in kiloparsecs.

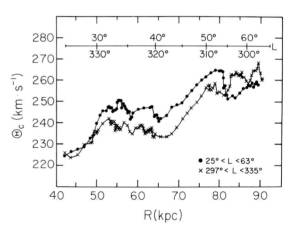

106. Kerr's two rotation curves. The galactic rotation curves for northern and southern sides of the galactic center have been derived from tangential 21-centimeter radio observations assuming circular rotation. The lack of smoothness of each curve and the differences between the curves are now being interpreted as caused by large-scale streaming associated with spiral features.

107. Radio spiral structure of our Galaxy (1967). This diagram was prepared by F. J. Kerr from 21-centimeter profiles observed in southern latitudes with the 210-foot radio telescope at Parkes, Australia (observations by Kerr and J. V. Hindman), and by A. P. Henderson from profiles observed from northern latitudes by G. Westerhout with the 300-foot radio telescope at Greenbank, West Virginia. The distribution in the inner parts is only roughly sketched in the diagram. The most noteworthy feature is the nearly circular pattern of spiral structure that emerges, with pitch angles of the order of 5°. The trough of low hydrogen density that can be traced from galactic longitude 20° to 80° is another remarkable feature. Regions of low hydrogen density are indicated by *L*. (Diagram from the University of Maryland.)

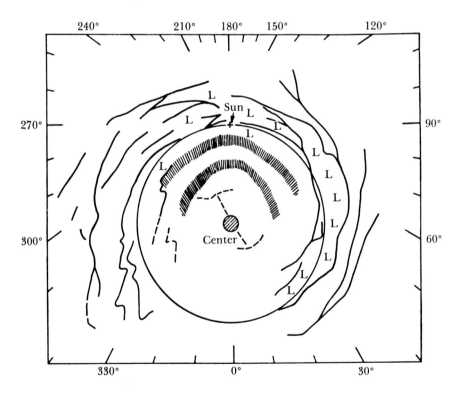

108. A preliminary map of radio spiral structure. Harold F. Weaver made this radio map on the basis of the Hat Creek 21-centimeter survey. It represents only the preliminary analysis (1972) of an extensive body of basic data. For comparison, Weaver has entered the optical data (see Fig. 103) of Becker and Schmidt-Kaler. The probable extensions of the observed features to the southern Milky Way (unobservable from Hat Creek Observatory, in California) have been sketched by Weaver.

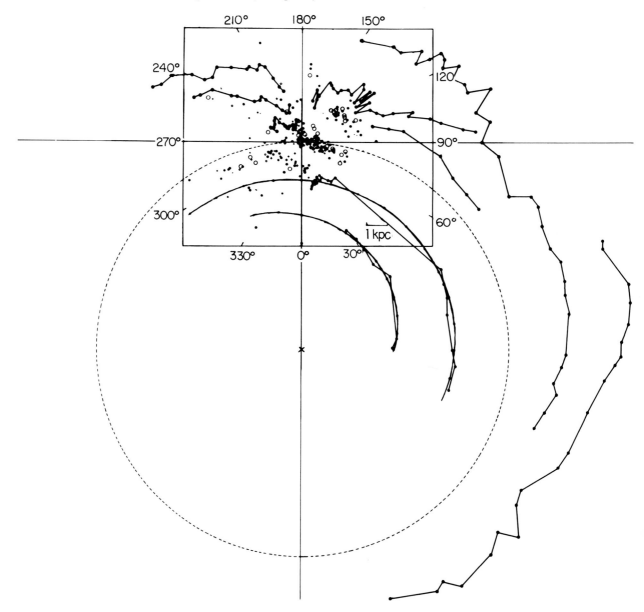

Spiral Structure in Neighboring Galaxies

The photographs of neighboring galaxies such as Messier 31 (Fig. 15), 33 (Fig. 109), 51 (Fig. 110), 81 (Fig. 111), and 101 (Fig. 16) show beautifully the sweep of the spiral structure of our nearest neighbors. There are, however, several properties known about these galaxies, which are recognized as the finest of spirals, that do not fit nicely into a straight and simple, all-inclusive spiral picture. Let us take Messier 31, the Great Spiral in Andromeda, as an example. First we reproduce the two famous photographs of it (Figs. 112 and 113) that Walter Baade made specially for the third edition (1957) of this book. In a study published in 1964, Arp reanalyzed some of Baade's material on the spiral structure of this galaxy. He rectified the photographs and drawings for a tilt of the central plane of the galaxy to the line of sight of 11°. In other words, he attempted to obtain a face-on view of Messier 31. His results are reproduced in Fig. 114. The Arp diagrams show that the emission nebulae found by Baade do give patterns that can be adjusted to logarithmic spirals, but we note from an inspection of the figure without the spirals that the basic pattern found by Arp has almost a closer resemblance to a ring structure than to true spirality. If we want to see spiral arms, we can certainly find them, and most of us would have no doubt that the basic structure in Messier 31 is of a spiral nature. And yet, the ring structure seems to have quite a bit to recommend it.

Several astronomers have attempted to trace the radio spiral structure from 21-centimeter profiles for the Andromeda spiral. The first study was made by Morton S. Roberts; more recently Rubin and Ford have again studied Messier 31 by combining the results of 21-centimeter and optical observations. There are marked H I peaks at the distances from the center where the Baade-Arp distribution charts of ionized hydrogen regions, observed optically, also show peaks. However, the highest concentration of neutral atomic hydrogen appears to be in a ring structure that mostly falls beyond the distances from the center of the Andromeda galaxy where the H II regions are found. This result bears an important similarity to one found from the analysis of hydrogen distribution in our Galaxy. It was shown some years ago by Westerhout that in our Galaxy the ionized-hydrogen distribution peaks at 5,000 parsecs from the center, whereas the neutral-hydrogen distribution peaks at distances of the order of 13,000 parsecs from the center. We find a similar situation in Messier 31.

A close examination of photographs of normal galaxies with spiral structure shows many worrisome features. In most of the beautiful spirals shown in the photographs of the *Hubble Atlas of Galaxies*, prepared by Sandage from Mount Wilson and Palomar photographs, the spiral arms are by no means smooth and continuous structures. There are often holes and bifurcations in them; loops and fringes occur in abundance. One may pity the astronomers who are placed inside some of the sections that are full of confusion and who may be attempting to unravel the details of spiral structure of the galaxies in which they find themselves. Hodge and others have traced the distribution of H II regions in several neighboring galaxies. On the whole, the spiral patterns stand out with reasonable clarity, but Hodge has found several galaxies in which there seems to be only confusion. Even with the best of optical and radio telescopes, the tasks of analysis for astronomers living and working in certain sections of

109. The spiral Messier 33 in Triangulum. The
photograph with the 200-inch Hale reflector
shows the central section of this spiral galaxy.
(Courtesy of Hale Observatories.)

110. The spiral galaxy Messier 51 in Canes Venatici. A photograph of the main spiral (NGC 5194) and its companion (NGC 5195), made with the 200-inch Hale reflector. (Courtesy of Hale Observatories.)

111. The spiral galaxy Messier 81 in Ursa Major. A
photograph of one of the most beautiful spiral galaxies, made with the 200-inch Hale reflector.
(Courtesy of Hale Observatories.)

112. Outer spiral structure of the Andromeda galaxy. The negative of a Mount Wilson photograph shows a section near the upper left-hand corner of Fig. 15. Dr. Walter Baade has given the following information about this photograph: Emission nebulae along one of the outer spiral arms of Messier 31, photographed in H-alpha light. At the bottom (center) emission nebulosities belonging to the next inner arm are seen, whereas Nos. 66, 67, and 1a are scattered members of one of the outermost arms of Messier 31. (Four-hour exposure at 100-inch telescope by Baade.)

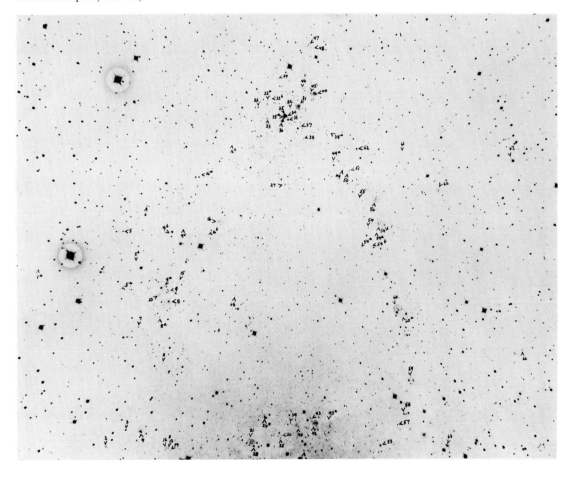

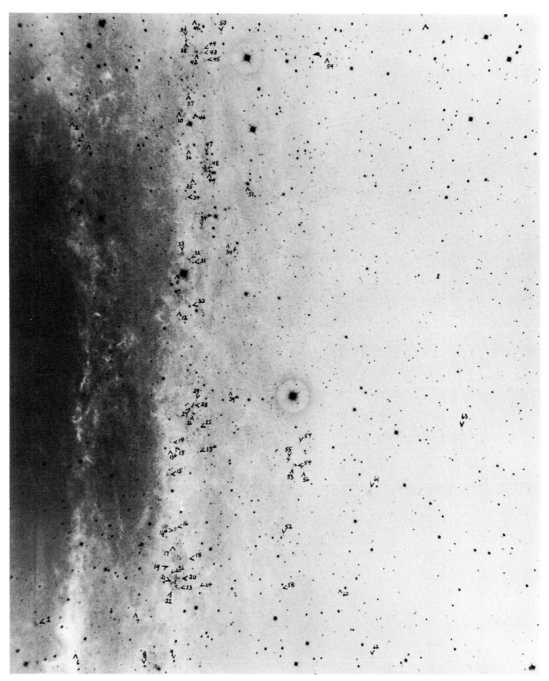

113. Inner spiral structure of the Andromeda galaxy. The negative of a Mount Wilson photograph shows a section to the right of center of the Andromeda galaxy as shown in Fig. 15. Dr. Walter Baade has given the following information about this photograph: Spiral arms of Messier 31, silhouetted against the bright nuclear region. Note that the emission nebulosities of the innermost spiral arm (Nos. 2, 3, 4, 5, and 6) lie in a dark lane; similarly, the emission nebulosities of the next spiral arm. This shows convincingly that the spiral arms are made up of large dust clouds which become conspicuous in this position because they cut off the light of the underlying Population II. (Four-hour exposure at 100-inch telescope by Baade.)

Messier 31, 33, 51, or 81 may be difficult. We would not be surprised if those astronomers in turn were to pity the poor fellows who live and work in the parts of our Galaxy not far from where we are located!

Messier 51 (Fig. 110) is one of the finest nearby spiral galaxies, seen almost face-on, with two well-developed arms that can be examined in great detail. Optically it shows very well how the most prominent spiral tracers, blue-white supergiant stars and their associated nebulosities, fix the over-all spiral pattern. Figure 115 shows side by side two photographs of Messier 51 and its companion. The photograph on the left displays beautifully the distribution of the hydrogen H-alpha concentrations. They outline very clearly the basic spiral pattern and help to define the manner in which one of the spiral arms merges into the companion. The negative picture on the right represents a normal blue photograph of the galaxy. These two photographs show that the spiral arms contain mostly gas and young stars, and that the luminous bridge connecting the main spiral and its companion has the same basic gaseous composition. Beverly Lynds has studied the distribution of obscuring matter in Messier 51 and she has concluded that the H II regions most frequently occur at the outer edges of the continuous dust lanes. Her diagram for the distribution of the dust is shown in Fig. 116. Messier 51 was one of the first to be studied with the Dutch radio interferometer, the Westerbork Array, which is a mile-long string of 12 interconnected radio telescopes. Figure 117 shows the result of the Dutch studies at a wavelength in the radio continuum near that of the 21-centimeter line, but well outside its limits. Mathewson, Van der Kruit, and Brouw find that the two strong continuous radio spiral arms coincide with the inner edges of the luminous optical arms. The radio spiral arms apparently follow very well the dust arms delineated by Beverly Lynds.

Figure 118 indicates that the spiral arms in Messier 51 are truly gaseous features. If the neutral atomic hydrogen were not actually concentrated to the spiral features in this galaxy, the 21-centimeter contours of equal intensity could be expected to wander about without reference to the optical spiral features; this obviously is not the case. One notes especially that the contours show the bridge connecting the main body (NGC 5194) and the companion (NGC 5195) as neatly as does the optical connection. To complete the picture for Messier 51, we should mention that Tully has made an optical study of the motions of the gas in the inner spiral regions. He has found some clearly defined streaming characteristics, which are in accordance with the predictions of the density-wave theory of Lin, Shu, and Yuan, to be described later in this chapter.

Though Messier 51 appears to be one of the prize spiral galaxies, it possesses one structural feature that is in a way very disturbing. This is the companion galaxy, which marks the end of one of the two major spiral arms. Arp has provided good evidence to show that the companion galaxy is truly associated with Messier 51 and he has furthermore suggested that it was probably ejected from the nucleus of Messier 51 as recently as 10 to 100 million years ago. He considers this observational evidence for a theory of Ambartsumian—which will also be described later in this chapter—according to which ejection of mass from the nucleus may be the source of all spiral structure.

The student of galactic spiral structure should always bear in mind the great variety

114. Emission nebulae in the Andromeda galaxy; two diagrams prepared by H. C. Arp. The left-hand diagram shows the positions of 688 emission nebulae (H II regions) in Messier 31, the Andromeda galaxy, as they would appear corrected for a tilt $i = 11°$ to the line of sight for the plane of M31. The diagram on the right shows the same plot with a logarithmic spiral fitted to the points. Note how difficult it is to distinguish between possible ring structures and spiral arms.

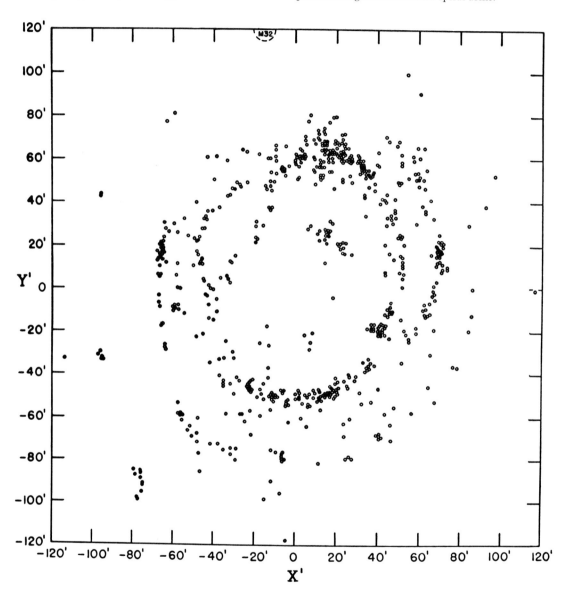

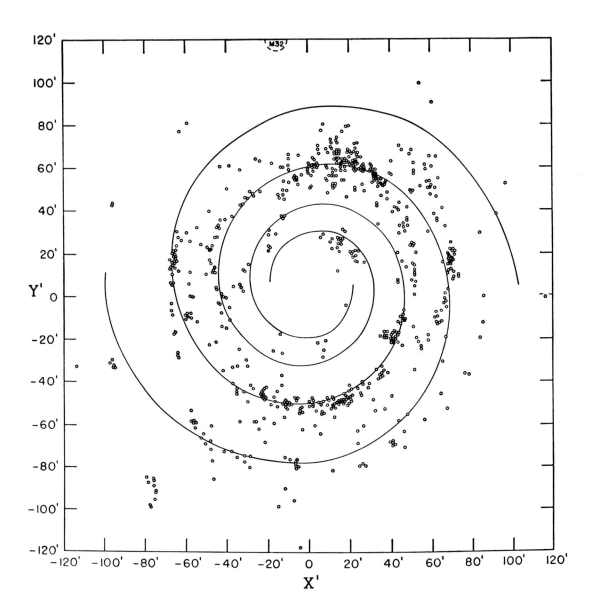

115. A photographic composite of two prints of
Messier 51, prepared by H. C. Arp of the Hale Observatories. The photograph on the left represents
a superposition of a negative print of M51, taken
with an H-alpha color filter with a pass band 100
angstrom units wide, and a positive print made
with a comparable filter centered at a wavelength
on the blue side of H-alpha. The result is a picture
of M51 in the light of H-alpha only. The photograph on the right is for comparison purposes; it
shows M51 in a negative print in normal blue
light. The H-alpha spiral features are shown beautifully in the photograph on the left. (Courtesy of
Astronomy and Astrophysics.)

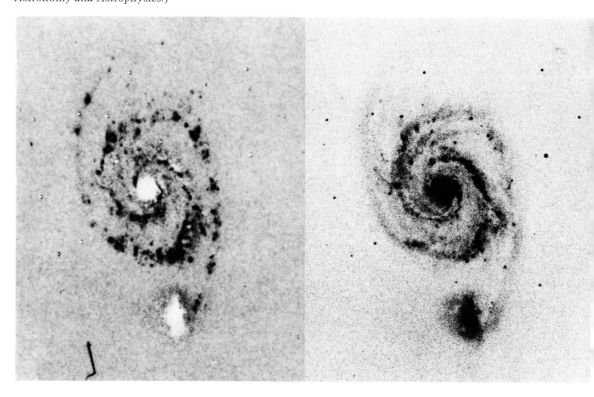

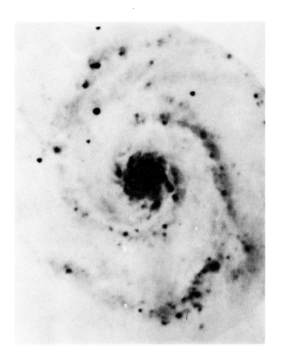
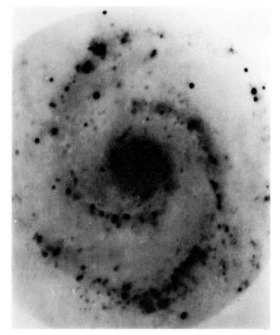

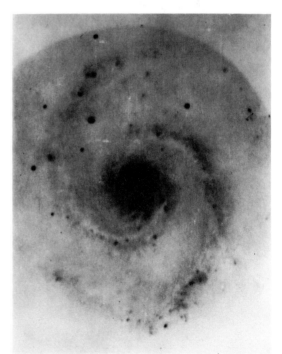

116. Obscuring matter in Messier 51. Beverly T. Lynds has prepared a series of photographs to illustrate the relation between the lanes of obscuring matter and the spiral arms in M51. The photographs at the top represent negatives of exposures through a broad-band filter in the blue (upper left) and through an H-alpha filter (upper right). The spiral arms are clearly shown. The photograph in the lower right is a negative of a plate taken with a narrow-band red filter (eliminating H-alpha) centered at 6,650 angstrom units. The drawing in the lower left is based upon one of Milton Humason's long-exposure photographs with the 200-inch Hale reflector; Dr. Lynds has drawn the dark lanes by proper shading and she shows the H II regions as black dots. (The photographs at the top were made with the 90-inch reflector of the Steward Observatory, University of Arizona.)

117. Radio emission from Messier 51. The ridges of radio-continuum radiation at 1,415 megahertz are shown superposed on Humason's photograph of M51 taken with the 200-inch Hale reflector. Dashed lines *A, B, C, D,* and *E* show interarm radio links, and *F* the link with the companion galaxy, NGC 5195. The ridge lines of radio emission coincide nicely with the absorption lines of Lynds shown in Fig. 116. (From a paper by D. S. Mathewson, P. C. van der Kruit, and W. N. Brouw.)

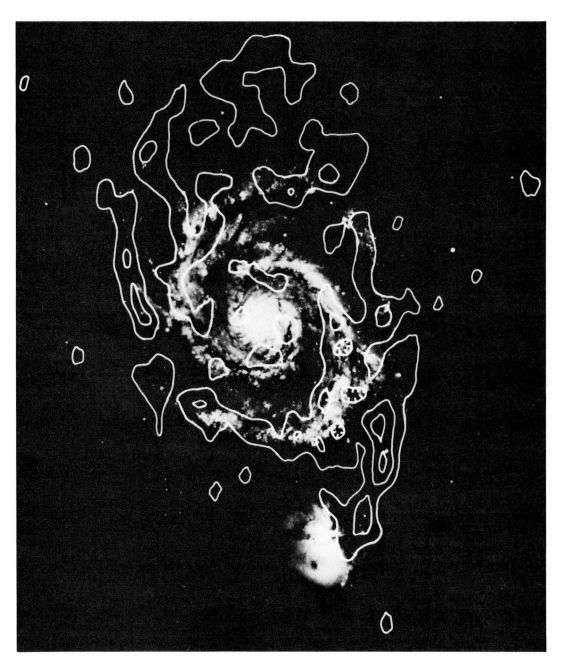

118. Equal-intensity contours for 21-centimeter radiation produced by neutral atomic hydrogen in the spiral galaxy Messier 51. The contours are shown superposed on a print of the 200-inch Hale reflector photograph reproduced in Fig. 115. Preliminary results with the Westerbork Array shown to B. J. Bok by Dr. Ernst Raimond (November 13, 1972).

of structures that occur in the spiral galaxies that have been photographed. Although it is true that the overriding property of spiral galaxies is that they show two trailing spiral arms, there are many objects that behave quite differently from this simple pattern. The nearby spiral Messier 101 (Fig. 16) is a multiple-arm spiral in its outer parts, and some of the more distant spirals show in their outer parts fireworks patterns with as many as six sections of arms or armlike features. Examples of complex patterns are shown in Figs. 119, 120, and 121. We should also bear in mind that there is a class of *barred spirals* (Fig. 122), objects with nearly circular spiral arms that seem to emanate from a central bar structure. The presence of an outer and inner ring structure appears to be quite common in spiral galaxies. Outer rings are shown beautifully in many of the barred spirals. Some astronomers have suggested that there may be evidence of a bar structure in our own Galaxy, but others—including the authors—have not accepted this suggestion.

Piddington of Australia especially has drawn attention to peculiar and interacting galaxies that do not fit into the basic two-arm spiral pattern that is shown in most of our illustrations of nearby galaxies. There are many examples of deviates in the *Hubble Atlas*. In 1959 B. A. Vorontsov-Velyaminov of Moscow published a remarkable *Atlas of Peculiar and Interacting Galaxies* and in 1966 Arp published his very useful *Atlas of Peculiar Galaxies*. Piddington has expressed the opinion that the exceptions may be the rule and that there is no basic two-arm design among spiral galaxies. He has attacked the assumption made by many, especially in the researches of one of the authors of this book, according to which spiral features are long connected streamers of hydrogen gas and young stars. Piddington doubts that there is any marked concentration of interstellar gas associated with spiral features. It seems as though the results obtained for Messier 51 with the Westerbork Array (see Fig. 118) refute Piddington's criticism and they prove conclusively that the H I spiral features appear as broad background features against which the material spiral arms are shown as narrow luminous traces. We note that very comparable conclusions may be drawn from the Westerbork 21-centimeter contours for Messier 81.

One of the most important lessons to be learned from an inspection of spiral galaxies is that the spiral phenomenon seems to possess a good deal of permanence. If it were a fleeting property of a galaxy, we would not expect to find so many of our neighbors showing spiral structure of a very similar type. If the spiral arms were in the habit of winding up, we would expect to see many more spiral galaxies with tightly wound inner spiral patterns than we seem to observe. The finest collection of galaxy photographs is contained in Sandage's *Hubble Atlas of Galaxies*. The Milky Way astronomer who is engaged upon an analysis of the spiral structure of our own Galaxy should plan to spend at least an evening each month just taking in the wonders of the *Hubble Atlas*. This is, in fact, a highly recommended activity for anyone interested in astronomy.

Spiral Tracers in Our Galaxy

O and B stars, singly, in clusters, or in associations, have traditionally proved to be about the most useful spiral tracers in our Galaxy. They are often found associated with H II regions, gaseous nebulae shining brightly in hydrogen H-alpha radiation. The stars responsible for exciting the H II regions are of

119. The spiral NGC 6946, photographed by J. B. Priser with the 61-inch astrometric reflector of the U. S. Naval Observatory Station at Flagstaff. Complex multiple structures prevail. (U. S. Naval Observatory photograph.)

120. The spiral NGC 5985, photographed by J. B. Priser with the 61-inch astrometric reflector of the U. S. Naval Observatory Station at Flagstaff. Again a very complex pattern of spiral features. (U. S. Naval Observatory photograph.)

121. The spiral NGC 5364 in Virgo. This photograph, made with the 200-inch Hale reflector, shows a spiral with remarkably wound-up continuous features. (Courtesy of Hale Observatories.)

122. The barred spiral NGC 1300 in Eridanus, from a photograph made with the 200-inch Hale reflector. (Courtesy of Hale Observatories.)

spectral class O, B0, B1, or B2, either single stars or clusters or associations of stars. A star with a spectral type between O and B2 is a blue-white giant or supergiant star. According to present evolutionary theories, such stars have been in existence for only a few, generally less than 10, million years. The O to B2 stars and the emission nebulae were considered to be most promising spiral tracers by Baade and by Morgan and his associates, and they continue to hold this place of honor today.

We described in Chapter 8 how radio-continuum radiation and radio alpha transitions contribute to our knowledge of the distribution and motions of the hydrogen gas. We noted there that the radio techniques used for these researches provide us with data on the distribution of ionized hydrogen. The alpha transitions yield, furthermore, very useful data on the radial velocities of the clouds under investigation. These two radio tracers therefore supplement very nicely the data on the distribution and motions of neutral atomic hydrogen obtained from studies of the 21-centimeter line.

Returning to the optical picture, we remind the reader that clouds of interstellar gas produce interstellar absorption lines in the spectra of distant stars. The best known of these absorption lines are the interstellar K line, produced by ionized calcium, and the interstellar D lines of neutral sodium. The most extensive work in this field so far has been done by G. Münch for stars within reach from northern latitudes; the southern Milky Way still awaits full exploration. High-dispersion spectra often show multiple interstellar absorption lines, which are presumably indicative of the presence of several gas clouds along the line of sight; for each observed component we can readily measure the radial velocity. These interstellar gas clouds are presumably concentrated in and along the spiral arms. The radial velocities found for the interstellar absorption lines make it possible to pinpoint their places of origin in very much the same way as we do with the gas concentrations found from 21-centimeter profiles and from alpha transitions. To derive the distance to individual clouds, we still require a velocity model for the Galaxy, which transforms these radial velocities for each particular direction in the galactic plane into distances from the Sun and distances from the galactic center. What we obtain basically in each case is kinematic information, that is, information about distributions of velocities. It is only by the use of a velocity model of our Galaxy that we can obtain the distances. However, any cloud that is detected through an interstellar absorption feature must lie nearer to the Sun than the star whose spectrum contains the feature.

The interplay between kinematic and structural observations promises to become increasingly important in future years. Very often we have extensive data for the spectra, magnitudes, and colors of the stars, star cluster, or association responsible for a given H II region. We can then derive the distance of the cluster or association on the basis of known absolute magnitudes and colors of embedded stars and from a knowledge of the estimated amount of intervening absorption. If we can measure as well the radial velocities of the stars and of the associated gas, we may proceed to check whether or not the observed radial velocities of the gas clouds agree with the values predicted from our model. A difference between the observed and the predicted velocities will suggest that there is something wrong with our basic velocity model of the Galaxy, or, alternatively, it will give an indication of local streaming. We await the time when we can say that our basic velocity model is reasonably well established, so that we may then use the observed differences as indicators of large-scale streamings. Recent 21-centimeter data as well as stellar and nebular optical radial velocities have shown in many places large-scale regional departures from circular motion, which suggest large-scale streaming of gas.

The increased emphasis on kinematic studies is an encouraging development. It represents an area on which we must concentrate in radio and in optical researches. We cannot stress too much the need for optical radial velocities for H II regions and their associated stars to complement radial velocities obtainable by radio-astronomical techniques. Without establishing any distance scales, we can find for certain directions in the galactic plane whether or not identical radial velocities are obtained from radio-astronomical and optical studies of the same gas clouds. In addition, we may compare radial velocities from gas clouds with those from associated stars or star clusters.

In recent years the results of many optical studies of radial velocities for H II regions have been published, most notably by Courtès, the Georgelins, Monnet, and Cruvellier of Marseille Observatory, and by J. S. Miller and M. G. Smith, associated with Kitt Peak National Observatory and Cerro Tololo Inter-American Observatory. Abundant radio hydrogen alpha radial velocities have been measured by Mezger, T. L. Wilson, Gardner, and Milne for the southern Milky Way and by Reifenstein, Burke, Altenhoff, Mezger, and Wilson for the northern Milky Way. All of this material is now being analyzed in conjunction with a rapidly growing body of stellar radial velocities. Especially useful in this respect are radial velocities for the longest-period cepheid variables and for red super-

giants, which have been accumulating in recent years through the work of Feast and Humphreys.

The cepheid variables with periods greater than 10 days are fair to good spiral tracers. They are supergiant stars that presumably have already gone through the O and B stages. They have probably also been red supergiants for cosmically brief intervals before entering the phase of long-period cepheid variability. Kraft has called attention to their potential as spiral tracers. He stresses that cepheid variable stars can be detected and studied to far greater distances than are within reach for O and B stars even with the largest reflectors. As of now, it seems very likely that the longest-period cepheids are the youngest ones and hence they seem to be the best potential spiral tracers.

Dark nebulae are as yet one of the least studied groups of objects, but the time has come to use them more extensively than in the past. Good distance estimates must be made for many dark nebulae seen along the band of the Milky Way. Most dark nebulae occur in very crowded regions, where the scale of Schmidt-telescope photographs is not sufficient for careful and detailed study, and hence we must look to the large reflectors for precision work on them. Large reflectors now under construction almost always have provisions for large-field wide-scale astronomical photography, and these instruments offer beautiful opportunities for further research on galactic dark nebulae. Photography with image-conversion tubes makes it possible to reach the faintest accessible stars in reasonably short exposure times. Intermediate- and narrow-band filters, now available for photoelectric and photographic work, should prove a great help in obtaining magnitude data and precise colors.

There are many useful special groups of

stars other than O and early B stars that make reasonably good spiral tracers. The B emission stars are, according to Th. Schmidt-Kaler, among the best of these. Lindsey Smith has shown that the hot emission-line Wolf-Rayet stars are also fairly good spiral tracers.

In all work relating to O and B stars, longest-period cepheids, and dark nebulae, we are beginning to pay attention to their relative distribution within specific spiral features. We have noted already that the dark nebulae are often found concentrated along the insides of the spiral arms and there are indications that the cepheids and the O and B stars trace arms or spiral features that do not coincide precisely with each other—nor, for that matter, with the 21-centimeter features. In the years to come, we must look for possible differences in details of distribution, which may be caused by age effects coupled with basically different kinematic properties. It is clear that we cannot afford to indulge in theoretical explanations and interpretations until we have a good observational basis for our speculations.

Supergiant stars of all varieties, from the blue-white O and B supergiants to the reddest M supergiants, are among the most useful spiral tracers. The discovery of such stars from objective-prism spectral plates for both the northern and the southern Milky Way is not a difficult task. The spectrum of any star of interest can now be photographed in intimate detail, because most of our larger spectrographs are fitted with image-conversion equipment, which makes it possible to obtain high-quality spectra of fairly high dispersion in relatively short exposure times. From these spectra one can make estimates of absolute magnitudes for the stars under investigation, and their radial velocities can be determined with precision. In recent years Humphreys has been especially active in this area. Her re-

sults have demonstrated that the supergiant stars may be the key to our penetrating to great distances for many sections in the band of the Milky Way. We now possess sufficient numbers of faint photoelectric standards to make it possible to measure *UBVRI* magnitudes and colors with precision for any star of interest. If desired, the same stars can be subjected to intermediate- and narrow-band photoelectric photometry.

It seems probable that galactic magnetic fields are not the major controlling factor in the production of the observed spiral patterns. Such fields must, however, have considerable organizing influence on details of spiral structure. The observed polarization effects in the light of distant stars prove that magnetic fields are associated with spiral features. The observed polarizations are generally largest, and the alignment of the polarization vectors appears to be most regular, when we view a star across a spiral arm; small percentage polarizations, combined with haphazard orientations of the vectors, are observed when we look longitudinally along an arm. Hence, present indications are that polarization effects are helpful in the tracing of spiral features, even though we are far from understanding just what roles large-scale magnetic fields play in theories of spiral structure.

We now have available extensive and beautifully precise material on the amounts and orientations of the polarization vectors for stars along the entire Milky Way, north as well as south (Chapter 9). The original data, which had been contributed by the discoverers of interstellar polarization, Hiltner and Hall, have been supplemented by extensive new catalogues by Mathewson of Mount Stromlo Observatory and by Klare and Neckel of Hamburg. They have published catalogues and charts with polarization data for the southern Milky Way, the sections for which

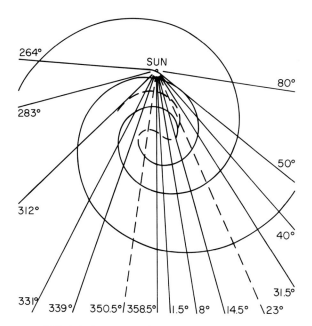

123. Spiral structure from radio edges. Anne J. Green of Sydney University has drawn this spiral diagram with a pitch angle of 12°.5 as fitting the steps in the radio-continuum observations for our Galaxy as well as several optical features. In this particular model, the Sun would be located in a spur.

such data had previously been lacking. Figure 100 shows the Mathewson polarization map.

Synchrotron radiation, which dominates at longer wavelengths in the radio range, has been helpful in defining certain edges in the longitudinal distribution of high-energy free electrons. These directions are identified as ones in which we look more or less length-wise along a spiral arm. At Sydney University, Mills and Mrs. Green have done the most effective research along these lines. They have located some marked jumps in the distribution of the strength of the integrated synchrotron radiation plotted as a function of galactic longitude. Mrs. Green's analysis of

the available data suggests that our Galaxy probably possesses a spiral pattern with average pitch angles in the range between 9° and 15°, but ring structures are not ruled out (see Fig. 123).

Pulsars and x-ray sources have been suggested as potential spiral tracers. Pulsars are probably the direct descendants of stars that have passed through the supernova stage, and supernova remnants are known to be the emitters of very powerful x-rays. However, the precise status of supernovae, pulsars, and strong x-ray sources as spiral tracers remains very much in doubt, at least for the present.

Gravitational and Other Theories

Ten to 15 years ago, the trend was to consider as the most likely hypothesis that galactic spiral structure is caused and maintained by large-scale magnetic fields, which were supposed to align the gas in spiral patterns. However, it was then discovered that the longitudinal magnetic fields average in all probability only 2 or 3 microgauss in strength, and they appeared too weak to produce the major controlling effects that are required. It is more likely that gravitational forces are responsible for the observed over-all spiral patterns and that the magnetic fields are frozen into these patterns as a by-product of the distribution of the ionized gas, mostly hydrogen (protons). Gravitational concentration of this ionized gas may locally produce magnetic fields of unusual strength. The behavioral patterns of the polarization vectors, such as those shown in Fig. 100, can thus be considered as a by-product caused by the concentration of ionized gas in spiral features.

Bertil Lindblad was for many years the lone champion of gravitational theories to explain spiral structure. In the middle 1960's C. C. Lin and his associates, Shu and Yuan, began a series of studies according to which the observed spiral features would be produced by a density wave passing through the interstellar gas. Lin starts out by assuming that the gravitational potential close to and in the central plane of a highly flattened rotating galaxy has a spiral component. He finds that gas and stars moving with near-circular velocities can, because of the effects of this spiral component, be subject to density waves, with the gas and the associated stars piling up for shorter or longer times in the gravitational spiral-shaped troughs of low potential. He investigates, with various types of spiral patterns for the gravitational potential, whether or not this piling up is a transitory phenomenon. He finds certain strict conditions that must be satisfied if the spiral potential field is to be a semipermanent one. If it is to persist, it must be sustained by the gravitational effects produced by the stars and gas that participate in the patterns of spiral behavior. Lin's theory predicts that in most highly flattened galaxies there will be a region in which spiral structure can prevail. For our own Galaxy this region lies roughly in the range of distances between 5,000 and 15,000 parsecs from the galactic center, which means that our Sun is, as it should be, right in the part of the Galaxy where spiral structure may be expected. Lin's theory predicts that the only stable and semipermanent spiral configurations will be those with two trailing spiral arms, emanating from opposite sides of the nucleus, and it also predicts quite precisely the spacing between the arms and their pitch angles. Spacings and pitch angles are determined by the nature of the underlying force field of the galaxy. The predicted pitch angles for our Galaxy are about $6°$, very much in line with the radio-astronomical results of Kerr, but out of line with the Becker-Morgan results. A typi-

cal calculated spacing between arms separates them by 3,000 parsecs and that is in fair average agreement with what we find in our Galaxy. Lin's theory predicts some large-scale streamings of gas and of young stars, with average stream velocities possibly as high as 6 to 10 kilometers per second relative to pure circular motion. Such deviations have actually been observed, and of the right order. The observed deviations from pure circular motion fit quite well into the pattern of the streamings predicted for our Galaxy by Lin's theory.

In Lin's theory, the spiral arms are produced by the spiral-shaped component of the field of gravitational potential. This component produces the density waves according to which gas and young stars are piled up along the troughs of minimum potential. In other words, the gas and young stars linger relatively longer in the spiral-potential grooves than in between, while the material in the spiral arm is being constantly replenished. In the earlier picture that most of us had (with the notable exception of Bertil Lindblad), the spiral arms were considered fixed loci of objects, such as clusters, associations, and emission nebulae. It seemed inevitable that, after one or two galactic revolutions, they should wind up, since the outer objects would move at slower speeds than those closer to the nucleus. This problem ceases to exist if one accepts Lin's view that spiral arms do not contain the same stars and gas indefinitely, that they are the product of density waves, and that the only semipermanent feature is the underlying spiral component of gravitational potential. We should note that the spiral pattern of gravitational potential in Lin's theory does not rotate as fast as the Galaxy for regions near the Sun. Hence we find that the instantaneous spiral arms do not persist, and the locus of maximum spiral features shifts with time and with respect to existing spiral arms.

There is as yet no good and complete theory to explain how the spiral potential fields of force come about in the first place. There is a slow rotation of the spiral-shaped field in the central plane of the galaxy; it resembles in some ways a two-armed boomerang! In the outer parts of the galactic spiral structure, the galaxy and the field rotate at about the same speed, but in the inner parts of a galaxy the boomerang does not move as fast as do the stars, the cosmic dust, and the interstellar gas that go around the galactic center with the speed of the general circular galactic rotation.

The stars are not bothered very much by the spiral potential field. However, in troughs of low potential the gas will have a tendency to pile up, slow down temporarily, and then move on to run again with the regular speed of galactic rotation, leaving the new gas behind it to go through the same performance. A spiral feature will be observed in the direction along the curve of greatest density, and in the course of several galactic rotations the gas will pass several times through these potential troughs, thus producing a steady spiral pattern of constantly varying gas clouds. In Lin's picture the spiral potential field, or boomerang, of our Galaxy moves at the Sun's position at a rate a little more than half that of the normal galactic rotation. Agreement between rotational speed and pattern speed is found at a distance of 14,000 to 15,000 parsecs from the center of the Galaxy. It will be noted that the shape of our boomerang fixes the shape of the spiral arms that are observed at any one time. Since the boomerang does not change its shape, the spiral arms should have the same appearance for as long as the density-wave pattern persists. There is no wind-ing-up problem in Lin's theory. It is worth noting here that the outer ring structures of neutral atomic hydrogen are found at just about the positions where the pattern moves at the normal rate of galactic rotation. Studies of the stability of the density-wave pattern have shown that a two-armed trailing spiral boomerang of potential has the best chance of surviving for several galactic revolutions.

The predictions of Lin's theory can be tested by mathematical calculations relating to the behavior of the density-wave pattern and they can also be checked by modern computer calculations for stellar systems with large numbers of gas clouds, or with hundreds of thousands of stars. Density-wave-like phenomena have been produced through computer calculations by P. O. Lindblad (Bertil's son) in Sweden and by R. H. Miller, Prendergast, Quirk, and others in the United States.

One of Lin's associates and former students, W. W. Roberts, has studied in some detail the processes that go on when the gas passes through the regions of lowest potential, that is, through the strongest part of the boomerang. He has found that sudden compression of the gas will take place and that along a wide front galactic shock waves will be produced. Narrow lanes where gas and interstellar dust are at very high density can be expected to be present. It seems that star formation might well occur under the expected conditions of high compression. The dust lanes would be at the inside of the spiral arm and the stars would migrate outward from it, with the youngest stars located closest to the shock front and the slightly older ones a bit farther removed. Star formation from interstellar gas and dust is a very complex phenomenon, and the passing of a density wave through the gas and dust cannot alone guarantee that star formation will take place on an extensive scale. However, if other conditions are right, then the passing of the density wave may well provide the required trigger action.

We should not leave the false impression that the Lin theory is the accepted gospel in spiral structure. It does represent a major breakthrough in a field where theory had made little headway. Its greatest asset to observing astronomers is that it presents them with firm predictions of the behavior, distribution, and kinematics of interstellar gas, cosmic dust, and young stars in and near spiral features.

The most severe of Lin's critics has been the Australian physicist Piddington, who, as we noted earlier in this chapter, favors a complete rejection of the density-wave theory of Lin, and especially of the shock-front hypothesis of W. W. Roberts, which is a part of the density-wave theory. Piddington has expressed the opinion that spiral arms are not truly gaseous features, but that they represent principally lines of concentration of young stars.

Piddington offers as a substitute for the approach of Lin and Lindblad a hydromagnetic theory of his own. In this theory the basic forces that are responsible for the observed spiral patterns arise through the encounter of our Galaxy (and others!) with a primordial intergalactic magnetic field. Our Galaxy's axis of rotation meets this field at an oblique angle. Piddington has shown that even a very weak intergalactic field will order the interstellar gas of our Galaxy sufficiently to produce in one galactic revolution patterns of concentrated regions of star birth resembling spiral arms. One interesting consequence of Piddington's theory is that it predicts spiral features away from the central disk of our

Galaxy. Such features have apparently been detected by R. D. Davies and by Verschuur. However, Piddington's spiral features are basically more like concentrations of recently formed cosmic dust and very young stars than like gaseous concentrations. In contrast to the Lin-Shu theory, which makes firm predictions about the velocity fields associated with spiral features in the central plane of our Galaxy, the Piddington theory predicts large velocities perpendicular to the galactic disk.

All workers in the field realize that there are many theoretical and observational approaches to the study of spiral structure. The density-wave theory does not exclude the presence of important effects produced by large-scale galactic magnetic fields, and it is quite reasonable that the importance of such fields be fully explored. Nuclear ejection, of the type favored by Ambartsumian and Arp, may provide a basic motive force for initiating and maintaining spiral structure. The researches of A. and J. Toomre have shown that encounters between galaxies may produce calculable spiral-structure-like phenomena. And, in the end, as has been stressed repeatedly by Pikelner of Moscow University, the barred spirals may well provide the major key to the understanding of spiral structure. There is obviously no such thing as a simple theory of spiral structure, and neither is it practicable to describe in a general way all of the complex phenomena shown by spiral features in our own and other galaxies.

Ambartsumian is one who has urged astronomers since the late 1950's to look toward the central regions of our own and other galaxies as providing the impetus for the spiral structure we observe in the outer parts. High-energy jets of ionized gas may enter the outer parts of a galaxy after having been generated by giant explosive phenomena in and near a galactic nucleus. The region of impact may be at 3,000 to 5,000 parsecs from the center, at which interface we may expect to find clouds of very fast-moving electrons entering the more peaceful sections of the galactic disk. It is at 5,000 parsecs from the center of our Galaxy that we find the great ring of giant H II regions noted by Mezger, Westerhout, and others. Large-scale magnetic fields may play more of a role than we have assigned them in our treatment.

Spiral Structure in Our Galaxy

It would be pleasant if we could offer for the end of this chapter a good diagram of the spiral structure of the Galaxy. But an inspection of Figs. 103, 107, and 108 shows that we cannot do so at the present time. We do not yet possess for the Galaxy the "Grand Design" for which C. C. Lin began to ask a decade ago.

Why are there differences between the Kerr-Westerhout and the Weaver diagrams of radio spiral structure? One reason for the discrepancies is that different interpretations are used by the analysts for the sections of the Milky Way in which sufficient 21-centimeter data are not yet available. Other differences arise because of variations in the method of analysis of the basic 21-centimeter profiles, and because of the use of differing velocity models for the Galaxy. One of the most important sources of discrepancies appears to lie in the manner in which authors connect the major concentrations of neutral atomic hydrogen into a single all-inclusive spiral pattern. The optical astronomer is faced with equally difficult problems in the analysis of his material. For example, most of us working in the field are agreed that there is a major concentration of young stars, gas, and dust in the direction of the southern constellation of Carina, and other spirallike features have been studied in Cygnus, in Orion, and in Norma. The Perseus Arm is well established over a considerable range of distance beyond the Sun and the Sagittarius Arm is equally well delineated at distances from the galactic center 2,000 parsecs less than that of the Sun. Becker and associates tend to interconnect these features into an over-all spiral pattern in which the arms have pitch angles of about 25°, whereas to Bok and others it has seemed more reasonable to connect these same features into a pattern showing several arms with pitch angles of 6° more or less.

The game of connecting recognized radio and optical spiral features into an over-all spiral pattern for our Galaxy will undoubtedly continue in the years to come. In the end, the basic pattern will probably be revealed by analyses of radio-astronomical data, especially 21-centimeter-line data. However, the analyses will by no means be straightforward. It is relatively easy to obtain definite results relating to concentrations of neutral atomic hydrogen in various directions of galactic longitude and latitude, but, for each direction, these concentrations will be observed as coming from clouds that have specific observed radial velocities of approach or recession. We shall require an intricate model of galactic rotation and of streaming motions applicable to the gas clouds in the Galaxy before we shall be able for each direction to translate the observed velocities into distances from the Sun.

To make life even more difficult, we encounter certain disturbing purely geometric effects in 21-centimeter analysis that complicate the situation. W. B. Burton and W. W. Shane have especially called attention to such effects. For some directions in the galactic plane, that is, at some galactic longitudes, the radial velocity of approach or recession will change very slowly with distance. This has a very disturbing consequence for the 21-centimeter profile at that galactic longitude. Even if the distribution of the neutral atomic hydrogen were perfectly uniform along the line of sight, there still would be an unduly large amount of that hydrogen in the direction that possesses a radial velocity close to the critical value. At the critical radial velocity the signal will build up in intensity, which produces an abnormally strong signal for that particular direction. The consequence of the very simple geometric effect is that we get a strong signal, which at first sight might wrongly be attributed to radiation from a cloud of neutral atomic hydrogen. *Velocity crowding* is one of the most serious concerns of the radio astronomer studying the distribution of neutral atomic hydrogen in our Galaxy.

We have noted that it is quite possible to isolate and study certain definite spiral features. At Steward Observatory, Bok, Humphreys, E. W. Miller, and others have in recent years made a comprehensive study of the distribution of gas, cosmic dust, and young stars for the Carina section of the Milky Way. The optical data alone, especially studies of O and B stars by Graham, show that in Carina we are probably looking edgewise along a major spiral feature. Graham showed that the sharp outer edge of this feature is at galactic longitude 283°, and there is evidence for some sort of inner edge near galactic longitude 300°. The feature has been traced optically over a distance of 7,000 parsecs, from 1,000 to at least 8,000 parsecs from the Sun. Figure 124 shows the diagram produced by Miller. The same sort of studies

have been made for a feature in Cygnus, where Dickel, Wendker, and Bieritz have studied the distribution and distances of 90 H II regions, all associated with a spiral feature stretching to at least 4,000 parsecs from the Sun in the direction of galactic longitude 75°. Sections of the Perseus Arm are equally well delineated, and the inner Sagittarius Arm is also clearly marked. The outlining and detailed studies of some of these features is exceedingly useful for an understanding of the properties of the spiral arms of the Galaxy. First of all, for a spiral feature like that observed in Carina, one may obtain the distribution of dust and gas and young stars of various ages across the feature, and thus hope to arrive at useful conclusions regarding the processes of star birth in spiral features. The Lin theory predicts certain systematic motions at the inner and outer edges of spiral arms. Observational evidence of deviations from pure circular motion associated with the Carina spiral feature has been obtained by Humphreys, who finds velocities of the order of 8 kilometers per second against galactic rotation on the inner side of the spiral feature and velocities of comparable amounts with galactic rotation on the outer side. This is of the amount and sense predicted by the Lin-Shu theory. For the Perseus Arm some very interesting phenomena have been found. Some of the gas clouds associated with the Perseus Arm appear to be rushing toward us at a rate of 20 to 30 kilometers per second, whereas the stars appear to move more or less in the circular orbits expected from basic galactic rotation.

We shall certainly continue our attempts to present diagrams showing the over-all spiral structure of the Galaxy and we shall of course try to obtain the Grand Design. However, in

124. A working diagram of the Carina spiral feature. Ellis W. Miller has prepared a diagram to show how the objects studied in the southern constellations of Vela, Carina, Crux, and Centaurus (galactic longitudes 270° to 300°) are concentrated to a curved spiral feature that can be traced between 2 and 8 kiloparsecs from the Sun. The interstellar obscuration reaches its greatest value on the inside of the feature, but there is marked obscuration on the outside as well. The center of the Galaxy is below the Sun, outside the diagram at a distance of 10 kiloparsecs from the Sun.

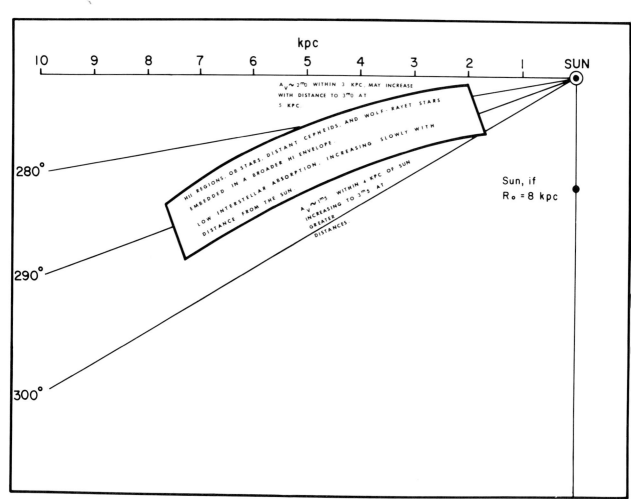

the meantime, much valuable research is being done on the detailed properties of isolated spiral features. The happiest aspect of present-day studies of galactic spiral structure is that the problems are being studied simultaneously from many angles. The theorists and users of large computers are hard at work on many aspects of the problem. Observing optical and radio astronomers consult and argue with one another uninterruptedly, and the astrophysicists are probing as best they can into the physical conditions of interstellar matter in spiral arms and in the regions between spiral arms. The full study of the spiral structure of the Galaxy is providing a stimulus for astronomical research on many fronts.

11
Our
Changing
Galaxy

Large-Scale Changes

Our Milky Way is changing constantly. The changes are slow and gradual and not easy to detect, for on the cosmic scale our span of life represents but a fleeting moment. Whereas the year serves as a convenient unit for time measurement on the Earth and for the recording of the changes in the positions of the planets, we shall have to turn to a larger unit in considering cosmic evolution. For this purpose we choose the *cosmic year*, which is the time of one revolution of our Sun around the center of the Galaxy. One cosmic year measures 250,000,000 of our familiar solar years.

What changes take place if we think in terms of cosmic years? The stars of the Galaxy are continually being reshuffled, for as small a difference in velocity as 1 kilometer per second will serve, in the course of 1,000,000 years ($\frac{1}{250}$ cosmic year), to separate two stars by 1

parsec. In the course of 1 cosmic year, groups of stars may be dissipated, new groups may be born, and the appearance of the Galaxy may undergo profound changes.

In addition to these purely mechanical changes, the physical make-up of our Milky Way system will not remain the same over intervals of the order of 1 cosmic year. In the deep interiors of the Sun and stars the nuclear energy sources are constantly at work, but they are not inexhaustible and the rates at which they produce radiant energy may in the long run vary considerably. The O and B stars use up their large yet limited supplies of nuclear energy at such prodigious rates that their lavish displays cannot persist for more than a fraction of a cosmic year. It seems almost certain that they were formed rather recently on the cosmic scale, that is, much less than 1 cosmic year ago. The intrinsically brightest supergiant stars have probable ages since birth of the order of 10 million years or less, which is a small fraction of a cosmic year. We are thus led to consider not only stellar evolution, but also processes of star birth. What role do the interstellar gas and dust play in the formation of new stars and to what extent do the stars replenish their not inexhaustible supplies of stellar matter? These are some of the questions that will concern us in this concluding chapter. First, we shall inquire into problems related to the appropriate time scales for cosmic evolution.

The fact that the individual stars are in rapid motion does not by itself imply that the system as a whole is changing appreciably. Red and white blood corpuscles are always racing through our veins and arteries, but the amount of blood in any particular vein does not as a whole vary greatly in the course of days or months. In the same fashion, the configuration of a knot in a spiral arm might remain the same over long periods of time; while some stars move out of a particular grouping, other stars may replace them.

If the distribution of positions and motions in a galaxy suffers no large-scale changes, we say that the stellar system is in *dynamic equilibrium*. Even if we should disregard real evolutionary changes—which we can certainly not do when we are dealing with the O and B stars—the state of dynamic equilibrium will not be a permanent one. For a particular group of stars, the state of dynamic equilibrium would probably survive for as long as 100 cosmic years, but not for 1,000 or 10,000 cosmic years. Chance encounters with passing stars will begin to have their effects over such long intervals of time and the state of dynamic equilibrium will be upset, to be replaced by the more thoroughly mixed state of *statistical equilibrium*, which is dynamically semipermanent. Detailed predictions for the future are not possible, but certain trends

may be demonstrated to exist and from them we may gain an insight into the time scale of dynamic evolution. We may, with considerable hesitation, attempt even to probe back into the dim past of the origin of the Galaxy.

When Stars Meet

How often, on the average, does a star come sufficiently close to our Sun to change the Sun's course appreciably? We know fairly well the average distances of separation and the velocities of the stars. We can compute without much difficulty how often a star will pass the Sun at a minimum distance less than that of Neptune. Such close approaches are rare; the Sun will, on the average, have such an encounter only once in 10,000 cosmic years, which means that it is unlikely that the Sun has suffered this sort of encounter with a passing neighbor during the 20 cosmic years of its existence. The total effect of the close passage probably would be rather minor and, although the event would be front-page news for a while, no permanent harm would be done. The orbits of the planets around the Sun would be changed, especially those of the outer planets, and Neptune or Pluto might join the interloper and leave the solar system. After the visit the Sun with its flock of remaining planets, and possibly with a few additions, would move in a direction that would deviate by about 20° from its original path.

What are the chances that the Sun might be hit by a passing neighbor? If that were to happen the results would be almost certainly catastrophic to life on Earth. But we probably have more immediate worries ahead of us, for the average interval between actual collisions of stars the size of the Sun is of the order of 1,000,000,000 (10^9) cosmic years or 250,000,000,000,000,000 (2.5×10^{17}) solar years! The only place where collisions might occur once in a while is in the central nucleus of the Galaxy, but even there the probability is not great.

The rarity of the phenomenon is best illustrated if we note that in our galactic system, with its mass equivalent to 100 billion (10^{11}) Suns, a collision between two stars should occur only once in 1 million years. If we consider that there are some regions of the Galaxy where the stars are closer together than in our neighborhood and that some stars have much larger target areas than our Sun, we might concede a somewhat higher probability for collisions. It is, however, very unlikely that even in a globular cluster, or in the dense nucleus of a galaxy, an actual collision will take place more frequently than once in 1,000 years.

From these simple computations it would seem that our chances of being put out of commission through a direct hit or close approach are very much less than the continual risk of being destroyed by comparatively small internal changes in the Sun. As small a persistent change in the Sun's total brightness as 1 magnitude would automatically result, in a fairly short interval of time, in a change of the average temperature of the earth by 75° centigrade. Such a change would not alter the planetary system as a whole, but it is improbable that life on the surface of the Earth could adjust itself to an average temperature around the boiling point of water or to conditions prevailing at 50° below zero centigrade. A mild variation in our Sun, something very much less drastic than a nova explosion, could easily put an end to all life on Earth.

But let us get back to our subject—stellar encounters. We have found already that spectacular close approaches are probably rather insignificant. We should, however, not forget

that the path of our Sun is continually being changed to some extent by more distant passages. A single star passing at a distance of 0.5 parsec will change the direction of the Sun's motion by somewhat less than 1 minute of arc. In the course of time the number of encounters within a few parsecs of our Sun is, however, quite large. Computations show that the cumulative effect of all encounters at minimum distances less than 3 parsecs between our Sun and other stars will, in the course of 1 cosmic year, be about equal to the total effect produced by a single encounter of the Sun and another star at a distance of 100 or 200 astronomical units. The greater importance of the cumulative effects of many unspectacular distant passages is demonstrated by the fact that a single encounter at a distance of 100 astronomical units should on the average happen only once in 2.5×10^{13} solar years or 100,000 cosmic years. It is quite clear that we might as well forget about the single "close" encounters and the head-on collisions and remember that the distant passages, because of their far greater frequency, are more effective in producing changes in direction and speed of the Sun's motion over long intervals.

Our Galaxy has not been rotating sufficiently long for the interchange of energy between stars of different types to have become effective. From our considerations of stellar encounters it seems very unlikely that the stars would still show so much individuality in their motions if our Galaxy had existed in its present form for much longer than 100 cosmic years = 2.5×10^{10} solar years.

The Evolution of a Star Cluster

Our part of the Galaxy is characterized by the presence of many loosely connected open clusters. Consider, for example, such clusters as the Hyades and the Pleiades. From our considerations of the disruptive effects of stellar encounters, these clusters would hardly seem very stable objects. The Hyades are relatively so close to us that we have not only accurate measurements of the motions of the brighter members, but also a rather complete census of the total membership of the cluster. The densest part of the cluster lies 40 parsecs from the Sun. The moving cluster contains about 150 members within a distance of 6 parsecs from its center. The speeds of the individual stars differ by not more than 0.5 kilometer per second from the average speed of the cluster as a whole.

We can predict what will probably happen to the Hyades cluster in the course of the next 10 or 20 cosmic years. The cluster ought to stay fairly close to the plane of our Galaxy and move therefore in a region where the stars are probably spaced much as they are in the vicinity of the Sun. We can easily compute how frequently a star that does not belong to the cluster, a field star, will pass through the cluster or how frequently one of them will pass by at a given distance.

Before we trace the effects of encounters on star clusters, we should first look a little more closely at the mechanics of a cluster free from intruders and passers-by. A cluster star that wanders away from the rest of the group will generally be pulled back by the attraction of the whole mass. The cluster tries desperately to preserve its unity. But there is a villain in the piece! We know from observations on galactic rotation that the stars in the vicinity of the Sun are all subject to the pull of the galactic nucleus. The parts of the cluster that are closest to the central nucleus of the Galaxy will feel more of the nuclear pull than those that are farther away. The general galactic forces will therefore tend to shear the cluster

apart. There results a regular tug of war. Can the force of attraction produced by the cluster as a whole pull hard enough to counterbalance the disrupting "tidal" forces from the galactic nucleus? If the answer is yes, the cluster will hold together, but if not, the cluster may soon be disrupted and its members strewn far and wide.

It takes some mathematical juggling to find where the dividing line will lie between clusters that can and those that cannot withstand the shearing forces of galactic rotation. A certain average critical density is computed. A cluster for which the average star density is less than the critical value cannot possibly stay together, but one with an average density above the critical value will generally have enough internal gravitational attraction to withstand the insidious disruptive forces from the galactic nucleus. A cluster for which the average density is equivalent to 10 solar masses for a cube 3 parsecs on each side should stay together if, in the course of time, it does not come much closer to the galactic nucleus than does our Sun.

I. R. King has made extensive studies of the dynamics of open and globular star clusters. He concludes that the dynamic future of a star cluster is generally established at the time of its formation from the gas and dust clouds of the Galaxy. These initial conditions determine the total number of stars of all varieties in the cluster and the degree to which the stars are initially packed; the radius of the young cluster is thus fixed. The stars move about inside the cluster with a distribution of velocities that settles down in a cosmically short time to a random Maxwellian type of distribution, and the main body of the cluster can stay that way for a long time. Matters are not so cheerful for the outer parts. Each stable star cluster has an outer boundary—the *tidal limit,* as King calls it— and the stars that lie beyond that limit will not be retained by the cluster, but will be torn from it by the differential shearing forces— the tidal forces—of the galactic nucleus. Only inside the tidal limit will the cluster exert enough total attraction to hold on to the member stars. King has developed methods for fixing the tidal limits rather precisely; they are in effect the boundaries of globular clusters. If the globular cluster moves in an elongated orbit around the galactic center (as most globular star clusters apparently do), the tidal limit is set at the position in the orbit where the maximum galactic tidal forces occur, which is at the position in the orbit of closest approach to the galactic nucleus.

What will be the probable effect of encounters between clusters and stars of the field? The dimensions of the clusters will in general be large and the passing field star will therefore not affect all cluster members in the same way. Two cluster stars that were originally pursuing strictly parallel paths will probably move after the encounter in slightly diverging orbits. The net result of each encounter will be to loosen up the cluster. This will tend to weaken the cluster's attraction on its members and the general galactic tidal force may then get a chance to do its disruptive work.

The Hyades cluster is probably dynamically safe for at least 4 more cosmic years or for 1,000,000,000 solar years. By that time the encounters with field stars will have done enough preliminary softening up to bring the cluster to the brink of disruption. About 5 cosmic years hence the galactic tidal force will become strongly effective. We may as well predict that 2,000,000,000 years from now we shall look in vain for the remains of the Hyades cluster wherever we may search.

The remaining aging stars will be scattered far and wide!

We now turn to the more compact clusters, such as the Pleiades and Praesepe, whose average densities are about ten times that of the Hyades. These clusters are in a far better state than the Hyades to withstand the disruptive shearing forces of galactic rotation; the time scale for disruption from this cause alone is probably of the order of 50 cosmic years. But here a different process is at work: the denser Pleiades and Praesepe will probably suffer a gradual internal collapse as a result of the escape of many members.

The process of collapse was first suggested in 1937 by Ambartsumian and independently in 1940 by Spitzer. It operates as follows: in a fairly dense cluster encounters between two cluster members will be considerably more frequent than encounters between a cluster member and a field star. On occasion a cluster member involved in an encounter with one of its fellow members may acquire a velocity large enough to permit it to escape—to "evaporate"—from the cluster. The escaping star will carry with it more than its fair share of the available total energy of the cluster, leaving the remaining stars with a somewhat reduced average energy. When there is less energy per star, the star members will not move quite so far from the center of the cluster as formerly. Hence the cluster will shrink. In the course of time many stars will escape, each of them taking with them more than their share of the total energy and, as the depletion progresses, a gradual collapse of the cluster will ensue.

The rate of escape will be slowed down somewhat, as Chandrasekhar has shown, because a *dynamic friction* is exerted by the slower-moving stars upon the escaping stars. However, the total effect, after correction for dynamic friction, still leads to the gradual dismemberment and slow collapse of clusters like the Pleiades and Praesepe. The predicted times in which purely internal collapse should take place are of the order of a few billion solar years—20 cosmic years, more or less.

The dynamic evolution of a star cluster will be affected by two additional processes:

(1) We have so far considered only encounters between stars. However, not infrequently a star cluster must pass close to—or through—a cloud of cosmic dust and gas. The masses of the known clouds of interstellar gas and dust run into thousands of solar masses, and, when a cluster of stars passes near one of these gas and dust complexes, there must be all sorts of disturbances at work. The gas and dust complex may produce far more disturbing tidal forces than those originating from stars or from the galactic nucleus. The life of the cluster may be severely shortened by a close passage involving it and a massive interstellar cloud. The times involved in such disruptions may be cosmically brief, but the effects should be startlingly disruptive.

(2) The second process that may affect the dynamic history of a star cluster is another wholly internal physical one. More and more, we are seeing evidence that stars lose mass by escape of gases from their atmospheres, especially in the later stages of stellar evolution. A large percentage loss of mass will make a star more susceptible to encounter effects. The lighter-weight star may pick up enough energy of motion from its neighbors and from passers-by to fly away from its parent cluster.

From a purely dynamic point of view, the time scale for important changes in the open clusters of the Galaxy seems to be of the order of 5 cosmic years. At the moment the Hyades, the Pleiades, and Praesepe are among the more conspicuous local features of our Galaxy. If we were to return in 10 cosmic years, these clusters would either have evaporated or collapsed—and there seem to be no others slated to take their place. One might be tempted to think about dismembered globular clusters as possible future Pleiades-like clusters, but two considerations show how impossible this would be. In the first place, globular clusters, with their characteristic spectrum-luminosity diagrams, cannot change into clusters with Pleiades-like color-magnitude arrays. Second, calculations readily show that the rate of evaporation for globular clusters is far too slow to lead to major changes through escapes in 10 or even 100 cosmic years. Dynamically, the globular clusters are among the most stable features of our Galaxy.

Might it be possible that other clusters not unlike the Hyades and Pleiades are now being formed from field stars? From a purely mechanical point of view it seems very unlikely that a workable mechanism could be found for the building up of clusters by chance encounters of unattached stars. However, it seems probable that new open clusters are being formed steadily by the process of the breaking up of clouds of interstellar gas and dust, followed by the collapse of the separate small clouds, or globules, into the stars that are the members of the newborn clusters. Such processes are rather likely to occur. It has been suggested by McCrea and others that there may be a compensating process of formation and of evolutionary decline of open clusters and associations of O and B stars, with the number of clusters and associations presumably not varying much in the course of time. For those open clusters with stars of spectral types A and later, the rate of depletion seems faster than the rate of formation of new clusters, and these may be a vanishing species.

We have, of course, no information concerning the numbers of open clusters that there were in the sky at the time when the first cockroaches began to march over the face of the Earth, a time which, according to the paleontologists, dates 1 or 2 cosmic years back. We know even less about the make-up of our Galaxy at the time of the birth of our Earth—an event that happened presumably some 20 cosmic years ago. The rate of disappearance of the open clusters that we know today is so high that it would seem extremely unlikely that the Galaxy could have existed in its present form for much longer than 50 cosmic years.

The Expanding Universe and the Cosmic Time Scale

Before we deal further with the physical problems of star birth, ages, and evolution, we should inquire whether there is not some reasonably well-established upper limit to the age of the universe of galaxies and to the ages of the stars in these galaxies. Such a limit is indeed set by the probable age of the expanding universe of galaxies. In the present volume we are primarily concerned with our Home Galaxy, and we must refer the reader to the Shapley-Hodge volume *Galaxies* (1972) for a broad discussion of the expansion of our universe and the related cosmic time scale.

Indications are that the interval between the beginnings of the expanding universe and the present covers 10 to 15 billion years. This total age is just about of the right order for the dynamic evolution of our Galaxy to fit into the scheme of things. Sixty years ago, V. M. Slipher showed from his studies at Lowell Observatory in Flagstaff, Arizona, that some

of the fainter galaxies, which he called nebulae, exhibit remarkable redshifts in their spectra. If these redshifts were interpreted as shifts caused by radial velocities of recession, then, so Slipher reasoned, several of his nebulae possess velocities of recession with respect to our Sun in excess of 1,000 kilometers per second. Around 1930, when it was recognized that Slipher's nebulae were true galaxies, Hubble and Humason of Mount Wilson Observatory extended Slipher's measurements to fainter galaxies. Since they could obtain approximate distances for these galaxies, they could study the universality of a redshift-versus-distance relation, which emerged from these studies.

Since the days in which Hubble and Humason did their basic work, the distance scale for galaxies has been adjusted considerably. Recent studies by Sandage and others, mostly based on data gathered with the 200-inch Hale reflector, show that the redshift-versus-distance relation is very nearly a linear one. If we assume that the redshifts are indicative of the presence of radial velocities of recession, which practically everyone accepts nowadays, then the redshift-versus-distance law becomes a fundamental one relating velocity of recession and distance. The whole of the universe of galaxies is apparently expanding; the rate of expansion is set by the fact that two galaxies 10 million parsecs apart seem to be receding from each other at a rate of about 550 kilometers per second, the value derived by Sandage and Tammann. Redshifts corresponding to radial velocities equal to half the velocity of light have been observed for normal galaxies, and redshifts corresponding to velocities of recession in excess of 80 percent of the velocity of light have been found for some very distant quasistellar sources, known as *quasars*. Hence the general expansion of the universe appears to be well established on the large scale. If the above expansion rate has more or less persisted in the past, then very simple calculations show that all galaxies participating in the expansion should have been very close together 17 billion years ago; this "age" suits the Milky Way astronomer very well indeed.

The foregoing calculation is of course a very rough one; for more precise results, one has to fit the observational data into a relativistic model of the universe of galaxies. We may summarize the situation by saying that the ages derived for various models of the expanding universe all come pretty close to our figure. The expansion does not need to have been a steady one. It may well be, for example, that our universe of galaxies began as a Big Bang, resulting from some gigantic explosion, and that the very rapid expansion rate that may have existed shortly after the explosion has begun to slow down. It is obvious that the time since the explosion, as derived from presently observed expansion rates, would then have to be less than the figure of 17 billion years that we have quoted. Also, it may well be that our universe is an oscillating one, which is at present in a stage of expansion, but which ultimately may collapse again upon itself.

There are many observations that support the hypothesis of the expanding universe. The quasars are almost surely galaxies that we observe now as they were 5 billion years or more ago. They show by their numbers at great distances that there was much more activity in the universe of galaxies 5 to 10 billion years ago than there is at the present time. Another line of support for the hypothesis that a terrific cosmic explosion actually took place about 10 billion years ago comes from an observation by Penzias and Wilson,

as interpreted by Dicke. They observe the remains of the energy originally associated with the Big Band as a microwave background radiation with an effective temperature of 3°K, which is found over the whole sky. According to the best available observational evidence, we can now reach galaxies, distant quasars, to distances as great as 8 to 10 billion light-years, about 3 billion parsecs. Such observations imply that we can look back in time to objects now observed from the Sun and Earth as they were 8 to 10 billion years ago.

How was our Galaxy formed? An answer to this question can be given if we bear in mind that the oldest star clusters and individual stars are found at large distances from the central Milky Way plane in our Galaxy. This would seem to imply, according to a theory developed by Eggen, Lynden-Bell, and Sandage, that shortly after the Big Bang our Galaxy became a separate unit, a large near-spherical blob of gas. When condensation of some of the original gas into stars and star clusters began to take place, it occurred probably all through the blob. As time progressed, the gas began to be concentrated more and more toward the central Milky Way plane, which then acquired its present rotation. Younger stars and clusters were formed as the gas cloud became flatter and flatter, and we are now in the stage in which the central gas (and dust) cloud is remarkably thin. Star birth now seems to be confined entirely to the regions of cosmic gas and dust within a few hundred parsecs of the central Milky Way plane. According to this attractive picture, the oldest globular clusters and the oldest open clusters were formed first. Star birth and cluster birth have long ago ceased in the halo of our Galaxy. We are fortunate that these processes have not, however, ceased to operate near the central plane of the Galaxy, and we are even more fortunate that the Sun and Earth are in a position close to this central galactic plane, and in the outskirts of our Galaxy, the parts of the Galaxy where the evolutionary pots are still boiling nicely!

Energy Production inside Stars

Our estimates of the cosmic time scale were based on data from stellar motions and the arrangement of stars in systems. What about the stars themselves? They ought to have an important voice in the matter, for they are asked to supply energy for millions and millions of years. It is all well and good to talk in terms of 10 billion years, but can the stars keep shining for so long a time?

Other volumes in the series of Harvard Books on Astronomy, notably Lawrence H. Aller's *Atoms, Stars, and Nebulae* (revised edition, 1971), consider carefully the processes by which our Sun and other stars are continually supplied with energy for radiation. The familiar processes of chemical burning have no place in the scheme. It is so hot in stellar interiors that the atoms are stripped of most of their outer electrons. Transformations like those of molecular chemistry can take place in the atmospheres of some cool stars and in our laboratories on Earth, but not in stellar interiors. At the high temperatures prevailing inside the stars, physics looks for the source of stellar energy in the changes that must take place in the atomic nuclei. For our Sun, the interior temperatures rise steadily from 6,000°K at the surface to some 15 million degrees in the central region.

At these tremendously high central temperatures the atomic nuclei, freed from most of their customary neutralizing entourage of electrons, move at such speeds that violent changes occur when two of them collide. Hydrogen nuclei—the familiar protons—collide with other hydrogen nuclei. These nuclear collisions lead to the formation of heavier nuclei, notably of helium; in some stars this comes about through a transformation cycle in which the carbon nucleus figures prominently as a catalytic agent, in others through direct interaction of protons with other protons. We refer the reader to the Aller book to learn precisely how these nuclear transformations operate, but the important fact is that two protons and two neutrons with a total mass of 4.033 atomic mass units must combine to form a helium nucleus, whose mass is 4.004 on the same scale. A small fraction of the original mass—about 0.7 percent—is therefore lost in the course of the nuclear construction. This mass is transformed into radiative energy according to the famous Einstein equation,

$$E = mc^2,$$

which relates the mass loss m to the corresponding released energy E, where c represents the velocity of light. The transformation of hydrogen into helium is the most effective process of energy generation at work in stellar interiors. Stars that have exhausted their plentiful original supply of hydrogen have some secondary nuclear processes on which to call, but none are so effective in releasing free radiational energy. Annihilation of matter at very high temperatures would presumably be a still more effective process, but we do not seem to find generally in the universe temperatures high enough for this process to occur.

Let us see how well the stars can get along in their evolution if they depend entirely on the energy produced by the transformation of hydrogen into helium. Our Sun, an average dwarf G star, radiates at a rather low rate; it should use up about 1 percent of its mass of hydrogen in 1,000,000,000 years, that is, in about 4 cosmic years. Hence, with about 60 percent of its mass still hydrogen, the Sun has enough nuclear fuel left to shine as it does now for another 50 cosmic years or more. But the picture is not so cheerful when we turn to typical A stars, like Sirius, with absolute magnitudes 3 to 4 magnitudes brighter than the Sun. These stars send out between 15 and 40 times as much energy per unit time as does the Sun. The masses of A stars are no more than 2 or 3 solar masses, so that the energy that must be generated per unit mass is about 10 times as great as for the Sun. Such a star should use up 1 percent of its mass as hydrogen nuclear fuel within half a cosmic year or so and it can hardly have sent out energy at its present rate for as long as 20 cosmic years. Hence the A stars can barely have existed since the time of the birth of our Earth, and, if they were formed that long ago, they must be getting near the end of their available supplies of nuclear hydrogen fuel.

What about the O and B stars, which may shine with intrinsic brightnesses equivalent to 100,000 Suns and which have masses less than 100 times that of the Sun? Here the nuclear processes must be producing energy at rates 1,000 times as fast as in the Sun and the total supply of nuclear fuel should be exhausted in a fraction of a cosmic year. Hence the O and B stars must be of very recent origin, and we are forced to look—a not unpleasant task!—for places in our Milky Way system where the process of star birth may be observed today. Hoyle, Bondi, McCrea, and others have suggested that massive stars may replenish their exhausted energy sources by accretion, that is, by the capture of matter from interstellar space. This process is probably not a very effective one. In order to ac-

quire any appreciable quantity of matter by accretion, a star must be practically at rest relative to the interstellar medium, and the radiation pressure it exerts on the particles of the medium must not be sufficiently great to counterbalance the tendency of matter to fall into the star's atmosphere. Neither condition appears to be satisfied by the spendthrift O and B stars.

We turn back to the OB associations and aggregates that were described in Chapter 4. We found several of these associations to be expanding. When we retraced their development in time, we found that the expansion in at least four well-established cases must have begun only 0.1 cosmic year ago. The short maximum lifetimes for O and B stars and the expansion phenomena in OB associations both suggest that the associations and their component stars are cosmically of very recent origin. Some were apparently formed a fraction of a cosmic year ago—a very short time compared with the age of our Earth, which is 15 to 20 cosmic years!

It is noteworthy that the O and B stars—which are known to be first-rate spiral tracers (see Chapter 10)—occur especially in parts of the Milky Way where interstellar gas and dust are plentiful (Figs. 125 and 126). It is tempting to hunt for relatively small condensations of interstellar gas and dust and look upon these as possible protostars. The small globules that are often seen projected against the luminous peripheral edges of emission nebulae (Figs. 89 and 90) appear to come closest to the protostars for which we are searching. They are found especially in the turbulent regions of high gas pressure near the outer boundaries of emission nebulae, the H II regions. Inside a globule contraction presumably will take place, possibly at first through outside pressure effects, and then

125. The emission nebula Messier 8. This short exposure with the 90-inch reflector of the Steward Observatory shows the core of Messier 8 and to the left of it a cluster of young hot O and B stars. Star formation has been pretty well completed in the section of the cluster, but there is still much gas left in the core region of the nebula.

126. The emission nebula Messier 16. This photograph shows the young cluster and the nebula known as Messier 16. The intricate patterns of overlying dark nebulosity are suggestive of the presence of shock fronts and turbulent conditions. From a Lick Observatory photograph with the 120-inch reflector by N. U. Mayall. (Courtesy of Lick Observatory.)

through the self-gravitation of the globule, which may grow slowly in the meantime through the accretion of gas and dust. This contraction should continue until the particles inside the globule shatter in collisions to become molecules and then atoms. Finally, the atomic nuclei move sufficiently fast that nuclear transformations result from collisions. When nuclear transformations occur, the collapse should slow down and gradually come to a halt; a star would have been born.

Globules and Dark Nebulae

The search for and study of protostars is one of the most intriguing tasks facing the Milky Way astronomer of the 1970's. A protostar will likely be the result of the collapse, through gravity or otherwise, of an interstellar gas cloud, possibly of a cloud contaminated with cosmic dust. Such a collapse may produce at first a very extended and rarefied star, with a radius of a few hundred astronomical units. A star can be said to have been born once the object sends out detectable radiation. It may well be that protostars are formed by the combined processes of collapse and fragmentation of the original interstellar cloud. Such a process would naturally produce a cluster of protostars.

The most obvious candidates for protostars are the small dark globules, which we discussed in Chapter 9 and in the previous section. Their probable masses are in the range between 0.1 and 1 solar mass. We also discussed larger globules, the variety that is not infrequently seen on our Milky Way photographs off by themselves, where they appear as roundish star voids against a background rich in stars (Fig. 94). Their radii are in the range 0.1 to 1 parsec, with estimated masses between 2 or 3 and 100 solar masses. They seem predestined for ultimate collapse into

protostars. We also find some large unit dark nebulae, like the one near Rho Ophiuchi (Fig. 93), that have radii as great as 4 parsecs and most likely masses of the order of 2,000 solar masses. These may be slated for fragmented collapse into a cluster of protostars. Grasdalen and the Stroms may have discovered a young OB association embedded in dense cosmic dust right in the heart of the Rho Ophiuchi dark nebula.

It is encouraging to note that the future for the detailed study of globules, large and small, and of unit dark nebulae looks promising. Already observation has shown that molecular lines in the radio range are observed coming from the directions of some of the best-known unit dark nebulae. Heiles has studied the OH bands reaching us from the direction of the Rho Ophiuchi dark nebula, and he finds that temperatures of the order of 5° to 10°K prevail in the interior of this nebula. As our observational techniques are further developed, it may become possible to search for and study molecular radiation reaching us from the larger globules and from smaller unit dark nebulae. Studies of 21-centimeter radiation should produce data on the neutral-atomic-hydrogen content of these objects, which is probably small, and it is now possible to check on the presence of molecular hydrogen, which is plentiful. From intensity ratios of critical molecular lines, we are able to determine the velocities of turbulence associated with the cool gas. Before long, we should be able to present the theoretical astrophysicist with data on the total masses, composition, temperatures, and turbulent motions in many of these objects. There are further contributions that the observational astronomer is making. For example, in the Rho Ophiuchi cloud (Fig. 93) star counts by one of the authors have shown that there is an observable and well-established density gradient for the

127. The globule Barnard 335. The small and roundish dark nebula is seen projected against a rich star field. Its diameter measures 4 minutes of arc, which, at a not improbable distance of 300 parsecs, amounts to about one-third of a parsec, 70,000 astronomical units. Radio studies of the small dark cloud by Palmer, Rickard, Zuckerman, and Buhl show that this globule contains formaldehyde, H_2CO. No stars are observed that lie beyond the globule.

cosmic grains in the cloud. This particular cloud resembles in many ways a globular cluster of cosmic grains! Carrasco and the Stroms have made a comprehensive photometric study of the Rho Ophiuchi region. They find that the sizes of the cosmic grains in the cloud are greater than average for the interstellar medium. They ascribe this to depletion by sticking of heavy elements onto the grains. A reasonably strong magnetic field (10^{-4} gauss) appears to be associated with the cloud. They estimate that the cloud has existed as a unit for close to 1 million years.

We are now gathering increasing evidence to show that some of the larger nebulae are observed in the act of breaking up into many units. The finest example of such an event is seen in the Southern Coalsack, where Tapia has found that a fragmentation process is already in progress. The Coalsack is not a near-spherical ball of cosmic dust, acting as a single unit, and neither is it a thin sheet with a variable surface density for the grains. Instead it appears to be a conglomerate of smaller units, some of them remarkably dense already at the present stage, and presumably headed for ultimate collapse.

Star Formation and Spiral Arms

The stars commonly found associated with spiral features are without exception very young. As we have seen, typical ages since formation are in the range between 10 and 25 million years, well below 1 percent of the age of the Sun and Earth. It is therefore only natural to look for evidence of continuing star birth as a phenomenon related to spiral structure. We spoke earlier of the Lin-Shu-Yuan density-wave theory of spiral structure being the one that currently is favored by many workers in the field. W. W. Roberts has shown that a high-pressure shock wave accompanies each density wave. Therefore, if clouds of cosmic dust and gas of more than average density are present in the interstellar material, they may be compressed by the shock wave to five or ten times their original density, possibly beyond the critical stage necessary to form protostars. These shock waves may serve as a trigger mechanism that sets off the processes of star birth along a spiral arm. Roberts has shown that the conditions for the piling up of dust and gas are most favorable along the insides of the spiral arms, and here is precisely where we observe the early beginnings of star formation. One naturally expects to find protostars and newborn stars along the inner edges of a spiral arm; the spectacular O and B giant stars are the brilliant and slightly older objects that mark the central line of the spiral arm.

The distribution of interstellar gas and dust is far from uniform in our Galaxy. Hence, as the shock wave moves through the interstellar medium, it will inevitably pass through some regions of below-average density, regions where the compression of the shock wave is insufficient to produce conditions of collapse followed by the formation of a protostar. It is therefore quite understandable that along a spiral arm we do not find a smooth distribution of young clusters and associations of stars. We should expect to find voids between the obvious concentrations of protostars and young objects. This picture of star formation seems to have very strong observational backing; the various stages are as plainly visible to the eye as are plants in successive stages of development in our garden.

Supernova Explosions

Supernova explosions are among the most spectacular of celestial phenomena. In 1054, Chinese astronomers observed a gigantic supernova explosion, in the position where we now find the beautiful Crab Nebula (Fig. 83). There is a cinder left behind, which is observed as a radio and optical pulsar and which appears to be a totally collapsed neutron star, rotating on its axis in the incredibly short period of $\frac{1}{30}$ second. The Crab Nebula is at a distance of about 2,100 parsecs from our Sun and hence strictly an object belonging to our Milky Way System. During 1971 attention was drawn to another supernova pulsar, this one located about 45° south of the celestial equator in the constellation Vela. This explosion appears to have produced aftereffects detectable in the sky up to angles of 30° or more from the central pulsar. The Vela pulsar and its associated Gum Nebula (Fig. 84) are at about one-fifth the distance from the Sun to the Crab Nebula, 500 parsecs.

A supernova explosion must have two immediate effects on the surrounding interstellar medium. First, considerable amounts of gas enriched in heavy elements are added to the interstellar medium around the supernova; second, tremendous amounts of energy are transmitted to the surrounding interstellar medium in the form of explosive shock waves. The neutral atomic hydrogen that must have been present in the surrounding interstellar medium before the supernova outburst would obviously have become ionized by the outpouring of ultraviolet energy. It should be noted that the supply of fresh energy is by no means exhausted a few hundred or 1,000 years after the original outburst, for the tiny rapidly rotating neutron star continues to pour energy into the interstellar medium. We remind the reader that pulsars and supernovae are potentially most productive sources of cosmic-ray particles with very high energies, which help maintain the ionization of the surrounding medium.

One of the most striking optical features of the Gum Nebula is the presence of a remarkable filamentary structure, the sort of effect that one would expect from the passage of energetic shock waves through the surrounding interstellar medium (Fig. 84). It has been suggested that star formation may well have taken place, and may still be taking place, in the region of the Gum Nebula. Several stars of high luminosity, which are apparently young on the cosmic scale of time measurement, have been marked as possibly having originated at the same time as the star that later on produced the supernova pulsar; some of these are typical "runaway" stars. The gaseous filaments now visible on our photographs must represent highly condensed gas, and it would not be surprising if some of these filaments were to break up into strings of young stars, or protostars. Many workers in the field consider it likely that the supernova phenomenon may be conducive to the formation of protostars, but it is a bit uncomfortable that we do not see any sort of mass production of protostars taking place before our eyes and neither do we find near supernova remnants an abundance of cosmically young O and B stars.

It would be naive to look for an abundance of protostars in the regions close to recent supernovae. The collapse of a gas cloud into recognizable protostars or young massive stars, singly, in clusters, or in associations, is a process requiring at least 100,000 years, in most cases probably as long as 1 to 10 million years. The Crab supernova explosion was observed in 1054, less than 1,000 years ago, and the supernova explosion at the heart of the Gum Nebula cannot have taken place much longer than about 30,000 years ago. Even the youngest observed hot stars in the region near the Gum Nebula must have antedated the supernova explosion by hundreds of thousands of years. The condensations from which these hot stars were born must have originated from explosive events that took place long before the recent supernova outburst that produced the existing pulsar. However, there is little doubt that recent supernova outbursts must have produced conditions favorable for future star formation in their surroundings. And young stars may still be formed today through the collapse of gas concentrations caused by supernovae of the past that are no longer detectable.

The Formation of Stars

Interstellar clouds have been studied by the radiation they emit in several quite different regions of the electromagnetic spectrum: at x-ray, ultraviolet, visible, infrared, and radio wavelengths. Such studies have yielded basic information on the properties of clouds, which are presumably on their way to becoming protostars. A variety of dark clouds and infrared objects have been discovered and classified. On the basis of such information we can ask: What kinds of mechanism are responsible for the development of a protostar and its collapse into a star?

In a recent survey of theories of the formation of stars, McNally of the University of London lists several processes that may be at work; in the end he favors star formation by collapse, with gravity as the major cause. His conclusions are generally confirmed by the studies of other astrophysicists, notably Hayashi and colleagues of Japan. Larson of Yale University has drawn special attention to a process by which one of two things could happen: either a cloud will collapse into many different units and form a cluster of stars, or it will collapse much faster near the center than in its outer parts. The second al-

ternative means that a star would be formed mostly from the material near the center of the cloud, and that the young star would be embedded in a large envelope of dust and gas. Such a star would have a truly murky atmosphere! Much attention is being given to the manner in which the extended atmosphere might collapse. It seems likely that it would rotate and that this rotation would play an important part in holding it up for some time. In one way or another the protostar must get rid of the angular momentum that is stored in the rotating cloud of gas and dust. It can do so most readily by forming dusty shells around itself and these in turn may well break up into planets. The theory of the formation of a protostar from a cloud of gas and dust seems to lead almost naturally to the formation of a planetary system.

Recent investigations of infrared objects strongly support the theory that stars are formed from collapsing clouds of gas and dust. Becklin and Neugebauer of the California Institute of Technology have discovered an infrared point source near the heart of the great nebula in Orion that is almost surely a very young star. Low and Kleinmann of the University of Arizona have found a second object close to the same region. This infrared source seems to be a compact dust nebula, probably with a newborn star or cluster of stars near its center.

Evidence of nebulae enclosed in dust shells has been forthcoming from radio-astronomical observations as well. Mezger and his colleagues have found a number of emission nebulae that emit strongly at radio wavelengths but are not detectable at visual wavelengths. The hypothesis is that these are "cocoon nebulae": brilliant nebulae embedded in clouds of interstellar grains. Their radiation in the radio region can pass through the surrounding

128. Evolutionary tracks of pre-main-sequence stars. A diagram prepared by C. Hayashi to illustrate the development of pre-main-sequence stars with masses between 0.05 and 4 solar masses contracting to the main sequence. Vertically he plots the logarithm of the intrinsic luminosity of the star in terms of the sun's luminosity and horizontally the logarithm of the surface temperature of the star. (From *Annual Review of Astronomy and Astrophysics*, 4:189, 1966. Copyright by Annual Reviews, 1966. All rights reserved.)

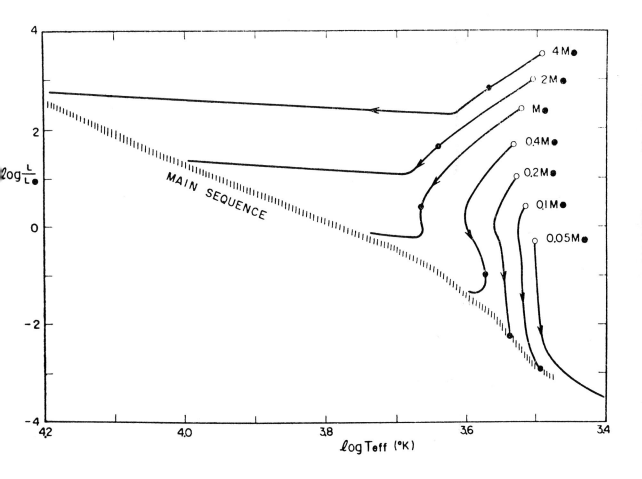

129. Nebulosity in Monoceros. This cometary nebulosity contains a dark globule near the head of the cometary structure. The globule measures about 1 minute of arc in diameter, which at the distance of the associated nebula NGC 2264 amounts to about one-quarter of a parsec, 50,000 astronomical units. There is a strong infrared source, discovered by Allen, directly above the globule. Zuckerman, Turner, Palmer, and Morris have found formaldehyde, H_2CO, in this region. From a photograph in red light with the 200-inch Hale reflector. (Courtesy of the Hale Observatories.)

dust clouds, but that in the visible region cannot. Some might be observable in the infrared.

A class of intrinsically faint stars, the T Tauri stars, almost certainly represents a very early stage of stellar evolution. T Tauri stars vary irregularly in their energy outputs, and their spectra show strong emission lines that are presumably produced in their extended outer atmospheres. The spectra of such stars also exhibit absorption lines that are formed deeper in the atmosphere. The lines are broad and fuzzy, indicating that the stars are either rotating rapidly or continually ejecting mass. There is much evidence to support the hypothesis that gases are steadily flowing out of the atmospheres of T Tauri stars. Such stars are most often found in groups, generally at the edges of or within dark nebulae. The fact that they cluster together is so marked that Ambartsumian gave them the name of *T associations*. The Mexican astronomer Mendoza has found that T Tauri stars are strong emitters of infrared radiation.

Herbig of Lick Observatory has apparently observed the formation of one truly new star, FU Orionis, which suddenly appeared on the scene in 1936. Quite recently Haro of Mexico has drawn attention to a star that behaves very much like FU Orionis: the faint variable star V 1057 in Cygnus. This star has recently flared up and is now very bright in the infrared. Haro expresses the opinion that FU Orionis and V 1057 Cygni were originally T Tauri stars and that now they have advanced to the next evolutionary stage, which is represented by a characteristic long-term flare-up. Herbig and Haro have discovered a number of small bright nebulae (referred to as Herbig-Haro objects) that are interspersed with the edges of dark nebulae, mostly in regions where T Tauri stars are abundant. T Tauri itself, the prototype for which the class is named, is embedded in such a nebula.

The general scheme we have described here is one that favors the formation of protostars through the process of gravitational collapse in clouds of interstellar gas and dust. We should point out that not every astronomer and astrophysicist favors this scheme. Layzer of Harvard has developed the following theory of the related formation of stars and galaxies. In the beginning the mass of the universe was distributed quite irregularly. Fragmentation and clustering took place on a large scale, and there were some blobs of hot, dense plasma (ionized gas) in which the gravitational field was much stronger than average and in which there were also large electric fields. Layzer considers conditions in these blobs as conducive to star formation. Another position has been taken by Ambartsumian. As we mentioned in Chapter 10, he considers it likely that violent explosions in the nuclei of galaxies (including the nucleus of our own Galaxy) may have much to do with the origin and maintenance of spiral structure. He suggests that associations of young stars along with interstellar gas and dust would be a direct result of the ejection of material by such explosions. So far neither his theory nor Layzer's has been developed to the point where it can be checked in detail by observations.

It seems quite natural that star birth should be occurring now in the spiral arms of our own and neighboring galaxies. Many dark nebulae and globules composed of interstellar gas and dust are seen almost in the act of collapsing into protostars or their close relatives. Objects that are either protostars or very young stars have also been observed. Infrared objects provide a natural link between small dark clouds and relatively normal stars. Quite a few of them may be cool, dense dust clouds with a star or a cluster of stars near the center. The cocoon nebulae may also supply new-born stars. The T Tauri stars seem to be the next stage and they help to bridge the gap between the protostars and the young stars.

Comprehensive research is continuing on the problems of change in our Galaxy and on the related questions of the birth of stars and their early evolution. It is a good thing to describe the Galaxy in all its majesty and to study the properties of its many components. The final aim, however, goes further. Deep in our hearts we want to know how the Galaxy came into being, how the stars were formed, and what is the future of the Milky Way system.

Index